CHEMICAL ENGINEERING METHODS AND TECHNOLOGY

FLUID PHASE BEHAVIOR OF SYSTEMS INVOLVING HIGH MOLECULAR WEIGHT COMPOUNDS AND SUPERCRITICAL FLUIDS

CHEMICAL ENGINEERING METHODS AND TECHNOLOGY

Handbook of Membrane Research: Properties, Performance and Applications
Stephan V. Gorley (Editor)
2009. ISBN: 978-1-60741-638-8

Non-Ionic Surfactants
Pierce L. Wendt and Demario S. Hoysted (Editors)
2009. ISBN: 978-1-60741-434-6

Handbook of Hydrogels: Properties, Preparation & Applications
David B. Stein (Editor)
2009. ISBN: 978-1-60741-702-6
2009. ISBN: 978-1-61668-167-8 (E-book)

Treatment of Tannery Effluents by Membrane Separation Technology
Sirshendu De, Chandan Das and Sunando DasGupta
2009. ISBN: 978-1-60741-836-8

Asphaltenes: Characterization, Properties and Applications
Jeremy A. Duncan (Editor)
2010. ISBN: 978-1-60741-453-7
2010.ISBN: 978-1-61668-515-7 (E-book)

Chemicals and Radioactive Materials and Human Development
Rajiv K. Sinha, Rohit Sinha and Shanu Sinha
2010. ISBN: 978-1-61668-145-6
2010. ISBN: 978-1-61668-455-6 (E-book)

Bulk Materials: Research, Technology and Applications
Teodor Frías and Ventura Maestas (Editors)
2010. ISBN: 978-1-60692-963-6

Chromatography: Types, Techniques and Methods
Toma J. Quintin (Editor)
2010. ISBN: 978-1-60876-316-0

Biomedical Chromatography
John T. Elwood (Editor)
2010. ISBN: 978-1-60741-291-5
2010. ISBN: 978-1-61668-470-9 (E-book)

Flame Retardants: Functions, Properties and Safety
Paulo B. Merlani
2010. ISBN: 978-1-60741-501-5

Bisphenol A and Phthalates: Uses, Health Effects and Environmental Risks
Bradley C. Vaughn (Editor)
2010. ISBN: 978-1-60741-701-9
2010. ISBN: 978-1-61668-514-0 (E-book)

Green Composites: Properties, Design and Life Cycle Assessment
François Willems and Pieter Moens (Editors)
2010. ISBN: 978-1-60741-301-1

Photocatalysis on Titania-coated Electrode-less Discharge Lamps
Vladimír Církva and Hana Žabová
2010. ISBN: 978-1-60876-842-4

Environmentally Harmonious Chemistry for the 21st Century
Masakazu Anpo and K. Mizuno (Editors)
2010. ISBN: 978-1-60876-428-0

Perovskites: Structure, Properties and Uses
Maxim Borowski (Editor)
2010. ISBN: 978-1-61668-525-6
2010. ISBN: 978-1-61668-870-7 (E-book)

Fourier Transform Infrared Spectroscopy: Developments, Techniques and Applications
Oliver J. Rees (Editor)
2010. ISBN: 978-1-61668-835-6

Fluid Phase Behavior of Systems Involving High Molecular Weight Compounds and Supercritical Fluids
Pedro F. Arce and Martín Aznar
2010. ISBN: 978-1-61668-310-8

X-Ray Photoelectron Spectroscopy
Johanna M. Wagner (Editor)
2010. ISBN: 978-1-61668-915-5
2010. ISBN: 978-1-61728-240-9 (E-book)

Preparation of Thin Film Pd Membranes for H2 Separation From Synthesis Gas and Detailed Design of a Permeability Testing Unit
M. Bientinesi and L. Petarca
2010. ISBN: 978-1-60876-538-6

Handbook of Research on Chemoinformatics and Chemical Engineering
A. K. Haghi (Editor)
2010. ISBN: 978-1-61668-504-1

Recent Progress in Chemistry and Chemical Engineering Research
A.K. Hagh (Editor)
2010. ISBN: 978-1-61668-501-0
2010. ISBN: 978-1-61728-237-9 (E-book)

Chemistry and Chemical Engineering Research Progress
A. K. Haghi (Editor)
2010. ISBN: 978-1-61668-503-4

Gadolinium: Compounds, Production and Applications
Caden C. Thompson
2010. ISBN: 978-1-61668-991-9
2010. ISBN: 978-1-61728-334-5 (E-book)

CHEMICAL ENGINEERING METHODS AND TECHNOLOGY

FLUID PHASE BEHAVIOR OF SYSTEMS INVOLVING HIGH MOLECULAR WEIGHT COMPOUNDS AND SUPERCRITICAL FLUIDS

PEDRO F. ARCE

AND

MARTÍN AZNAR

Nova Science Publishers, Inc.

New York

For permission to use material from this book please contact us:
Telephone 631-231-7269; Fax 631-231-8175
Web Site: http://www.novapublishers.com

LIBRARY OF CONGRESS CATALOGING-IN-PUBLICATION DATA

Arce, Pedro F.
Fluid phase behavior of systems involving high molecular weight compounds and supercritical fluids / Pedro F. Arce and Martmn Aznar.
p. cm.
Includes bibliographical references and index.
ISBN 978-1-61668-310-8 (softcover)
1. Phase rule and equilibrium. 2. Supercritical fluids. 3. Molecular weights. 4. High pressure (Technology) 5. Fluid dynamics. I. Aznar, Martmn. II. Title.
QD503.A73 2009
541'.363--dc22

2010001163

Published by Nova Science Publishers, Inc. ✢ New York

CONTENTS

Preface		**ix**
Chapter 1	Abstract	**1**
Chapter 2	Introduction	**3**
Chapter 3	Thermodynamic Models	**9**
Chapter 4	Phase Equilibrium at High-Pressures	**21**
Chapter 5	Results and Discussions	**25**
Chapter 6	Conclusions	**107**
References		**113**
Index		**123**

PREFACE

Although supercritical fluids (SCF) and their unique solvent characteristics have been a matter of continuing scientific interest since the past century, their potential benefit to chemical processing have not been fully realized; it is only in the past three decades that SCF solvents have been the focus of active research and development programs, especially in the area of polymer processing.

Chapter 1

ABSTRACT

Although supercritical fluids (SCF) and their unique solvent characteristics have been a matter of continuing scientific interest since the past century, their potential benefit to chemical processing have not been fully realized; it is only in the past three decades that SCF solvents have been the focus of active research and development programs, especially in the area of polymer processing. In general, the phase behavior of systems with high molecular weight compounds (polymers, copolymers or blends) is more complex than the one of systems with low molecular weight substances, since depends strongly on the energetic interactions and on size differences between polymer and solvent molecules. For instance, the understanding of the polymer + solvent interactions is of fundamental importance in the development of new products and processes in several important industrial sectors, such as special paints, cosmetics, packing, membranes, etc. Although the phase behavior of polymer solutions often exhibits a pronounced density dependence at high temperatures, and the nature of these physical interactions is well known, the thermodynamic modeling of these systems in industrial applications presents difficulties, mainly due to the necessity to characterize not only the polymeric phase, in terms of properties that are easily measurable, but also the interaction of the heavy components with the light components, through conveniently chosen parameters. For these reasons, it is important to use thermodynamic models that are able to describe the fluid phase behavior of polymer + supercritical solvent mixtures at pressure and temperature range used commonly in engineering processes. Several thermodynamic models or equations of state (EoS) have been used to describe the phase behavior of polymer solutions. As mentioned above, the incorporation of SCF solvent-based technology for polymer processing would be greatly facilitated if it were possible to simulate different process scenarios with an accurate EoS. A major objective of this modeling is to predict the changes in phase behavior observed as a function of solvent quality or as a function of polymer architecture with a minimum number of fitted parameters. There are several different EoS that can be used to calculate polymer-SCF solvent phase behavior; in this work, two non-cubic and one cubic EoS are used: the Sánchez-Lacombe (SL), the Perturbed-Chain Statistical Associating Fluid Theory (PC-SAFT), and the Peng-Robinson (PR) equations.

Chapter 2

INTRODUCTION

In recent years, many factors influencing phase behavior of high molecular weight compounds such as polymers, copolymers and blends have been studied because of their widely technological importance, as a simple method of formulating new materials with tailored properties which make them suitable for a variety of applications. On the other hand, systems involving high molecular compounds are more complex than systems with low molecular weight substances. Thus, the phase behavior of polymer solutions depends strongly on the energetic interactions and on size differences between polymer and solvent molecules; this phase behavior often exhibits pronounced density dependence at high temperatures, and even if the nature of these physical interactions are known, the thermodynamic modeling of the phase behavior of these systems in industrial applications presents difficulties, mainly for characterizing not only the polymeric phase, in terms of properties that are easily measurable, but also the interaction of the heavy components with the light components, through conveniently chosen parameters. It is important the understanding the phase behavior of polymer solutions in supercritical fluids (SCF) because it has great theoretical and practical interest. Predictions phase boundaries (e.g. cloud points) and phase compositions (e.g. solubilities) for such systems is difficult because they are highly non-ideal at high pressure, the polymer and solvent differ greatly in size, and finally, commercial polymers are composed of many molecules, differing in molar mass and chemical composition. The selection of SCF solvents to dissolve polymers is often challenging for processing applications because it is difficult to find a good SCF solvent that will dissolve the polymer at relatively moderate conditions. Carbon dioxide (CO_2) is the favorite solvent in SCF processes because it has a relatively low critical temperature and pressure and because it is inexpensive, nonflammable, non-toxic, and readily available. Polymers in general have very limited solubility in SCF CO_2 at temperatures below 80 °C (Hsiao et al., 1995; McHugh and Krukonis, 1994), although solubilities can increase significantly at higher temperatures for elevated pressures (Garg et al., 1994).

With the increase in the plastics production, plastic disposal has been regarded as a serious environmental problem; therefore, biodegradable polymers have received much attention in recent years as one of the approaches to solve the problem (Hoshino and Isono, 2002). A biodegradable polymer is a high molecular weight polymer that, owing to the action of micro- and/or macro-organisms or enzymes, degrades to lower molecular weight compounds (Karlsson and Albertson, 1998). Biodegradable polymers have received much attention as materials for reducing environmental problems caused by conventional plastic wastes; consequently, production of these materials has been studied and their commercial applications are growing progressively. Another engineering applicaton is the manufacture of polymeric foams, where compounds that contain chlorine and fluorine atoms (chlorofluorocarbons) have been used as physical blowing agents for a number of years, until they were banned in 1997 by the Montreal Protocol (Sato et al., 1996, 2000), because these gases are dangerous to the ozone layer; since then, hydrochlorofluorocarbons, hydrofluorocarbons, hydrocarbons and other gases, such as nitrogen and carbon dioxide, have become alternative blowing agents, since their ozone depletion potential is much lesser. Thus, the modeling of the solubility of these gases in polymers is very important for the optimal design of the foaming process. On the other hand, determination of solubilities of gases in molten polymers is another engineering application of considerable importance. Such data are of interest especially for the optimal design in polymer finishing processes; for instance, where the molten polymer is blanketed with inert gas; in certain specialized operations such as foam extrusion and fluidized-bed coatings; in the manufacture of polyvinyl chloride; and also when unreacted monomers, that are harmful to the environment, exist in the polymerization products and, therefore, a devolatilization process is needed to remove them. All these processes need information about gas solubilities in polymers at various temperatures and pressures (Durrill and Griskey, 1966; Peng et al., 2000). In order to find the conditions met in the process design and operations, molecular thermodynamic models, such as equations of state, can be used to describe the gas-liquid equilibria (Wang et al., 1994).

About biodegradable copolymers, researches on materials for biomedical applications have seen tremendous advances over the past 30 years; among these, the properties of these compounds make them ideally suited for orthopedic applications. The most used biodegradable materials are the poly(lactide) and poly(glycolide). In the 1960s, poly(D,L-lactide) (PLA) was proposed as a biocompatible, biodegradable and bioresorbable material for biomedical applications (Kulkarni et al., 1971). In recent years, environmental concerns have led to an escalated interest in PLA, as well as others biodegradable polymers, as an alternative to traditional commodity plastics (Mayer and Kaplan, 1994). PLA has the advantage of being not only biodegradable but also renewable, since the raw material, lactic acid, may be produced by microbial fermentation of biomass. PLA is a well known polymer for applications in the biomedical field. It has been used for more than 20 years for surgical devices such as sutures or clips. In recent times, these biomedical applications have been extended to controlled drug delivery systems and larger parts, such as screws for fracture fixation. PLA is highly accepted because of its good mechanical properties combined with its biocompatibility and its ability to degrade both in vivo and in vitro (Jacobsen et al., 1999). Poly(butylene succinate) (PBS) and poly(butylene succinate-co-adipate) (PBSA) are also biodegradable plastics produced from 1,4-butanediol, succinic acid and adipic acid as principal raw materials. PBS and PBSA are also known commercially as Bionelle, a biodegradable aliphatic polyester (Teeraphatpornchai et al., 2003). Imaizumi et al. (1999) found that PBS has viscoelastic properties and processability for direct extrusion gas foaming. Therefore, copolymers of glycolide with both L-lactide and D,L-lactide have been developed for both device and drug delivery applications (Middleton and Tipton, 2000). Copolyesters of PLAG are the most widely investigated polymers with regard to toxicological and clinical data (Witt et al., 2000) and have generated tremendous interest due to their excellent biocompability and biodegradability (Jain, 2000). Although these polymers have been commercially used for peptide delivery (ZoladexTM or DecapeptylTM), their utility for proteins is limited due to their degradation behavior and release properties (Domb et al., 1997). Using the poly(lactide) and poly(glycolide) properties as base materials, it is possible to copolymerize the two monomers to extend the range of homopolymer properties. For these reasons, it is important to know more about the location of the phase boundaries for PLAG + SCF mixtures in the industrial process production of this biodegradable copolymer.

During the last decades, several theoretical and experimental approaches were used to investigate the structure and thermodynamic properties of polymer and their blends, since that, in recent years, the use of polymer blends has been increased due to several reasons which have stimulated the interest to investigate physical blends of different polymers. For instance, for engineering applications, polymer blends are a cost-effective way to produce improved polymer properties; many times, polymer blends give the chances to understand and predict the relationship between structures and physical properties (Chang and Bae, 2003). On the other hand, the effect of solvent on polymer blends and block copolymers has important consequences for material processing, so much so numerous theoretical and experimental studies have demonstrated that the addition of solvents can either suppress or induce polymer phase separation in blends, depending on the relative quality of the solvent for each of the constituents (Rao and Watkins, 2000).

Block copolymers are receiving a lot of attention because they offer a great opportunity to the chemists to create mixtures involving polymer blends with desirable physical characteristics (Helfand and Wasserman, 1976; Lodge, 2003). Block copolymers consist of two or more polymer fragments chemically different, or blocks, linked covalently to form a larger and more complex macromolecule (Fasolka and Mayes, 2001). Block copolymers have an important paper as additive in many industrial products as paints, lubricants, coatings, etc, and are very strong candidates for potential applications in advanced technologies such as information storage, drug delivery, and photonic crystals (Urban and Takamura, 2002), due to their capacity to stabilize colloidal suspensions. This capacity is due to the way in that the block copolymer is adsorbed. Block copolymers (in the most simple case, diblock copolymers) are adsorbed onto a surface from a solution if one of the blocks has a high affinity for the surface (anchor block), while the other is standing in the solution (buoy block) (Fleer et al., 1993). The situation is similar, in some respects, to the case of terminally attached polymer chains (end-tethered chains) (Alexander, 1977; de Gennes, 1980); the main difference between these cases is that, since the anchoring block has a finite size (much greater that an anchoring point), it can influence the structure of the adsorbed layer by changing, for example, the distance between buoy blocks, depending on its size. This has important consequences for the interaction among the two layers of the adsorbed diblock copolymers, and the formation of bridges in this situation is unlikely. Consequently, diblock copolymers are better stabilizers than the homopolymers. Because of the oscillation free from their blocks, adsorbed diblock copolymers are interpreted and theoretically frequently modeled as a set of chains. For instance, Hadziioannou et al. (1986) applied the analysis of Alexander and de Gennes for sets of chains for adsorption of diblock copolymers (Evers et al., 1990).

On the other hand, phase behavior of polymer and copolymer solutions depends strongly on energetic interactions and on size differences between polymer or copolymer and solvent molecules. At higher temperatures (near to the solvent critical point), polymer precipitates due to the free volume effect. This type of phase transition is known as the Lower Critical Solution Temperature (LCST), which is characterized by the increase of pressure transition values with temperature. At lower temperatures, differences of energetic interactions between polymer and solvent molecules may lead to another phase transition (limited polymer solubility and phase split) known as the Upper Critical Solution Temperature (UCST) (Dariva et al., 2001). Figure 2.1 presents a typical pressure-temperature (PT) diagram for an amorphous polymer + solvent systems. If the asymmetry between polymer and solvent molecules increases, the LCST and UCST curves approach one of other and eventually, they can be mixed in a single denominated U-LCST curve (Folie and Radosz, 1995). For these reasons, it is important to know the location of the phase boundaries for polymer-solvent mixtures in industrial production of biodegradable polymers and copolymers. For polymer blends, a typical generic phase diagram of the UCST and LCST type is shown in Figure 2.2. There are three regions of different degree of miscibility: (1) single-phase miscible region between the two binodals; (2) four fragmented metastable regions between binodals and spinodals; and (3) two-phase separated "spinodal" regions of immiscibility, bordered by the spinodals. The diagram also shows two types of phase transitions which have been reported on the basis of the temperature dependence of the segmental interaction parameters (critical solution temperatures): the lower, LCST, the temperature above which two polymers phase separate (at higher temperature), and the upper, UCST, the temperature above which two polymers mix (at lower temperature) (Paul and Newman, 1978; Sánchez, 1982). The phase diagram with two critical points is a rule for mixtures with low molecular weight component(s), whereas the polymer blends usually show either LCST (most) or UCST (Utracki, 2002). A few blends having UCST are, for instance, PS with SBS, PoClS, PBrS, or poly(methyl-phenyl siloxane); BR with SBR; SAN with NBR; etc. (Utracki, 1989)[1].

1 PS: polystyrene, SBS: poly(styrene-co-butadiene-co-styrene), PoClS: poly(o-chlorostyrene), PBrS: polybromostyrene, BR: butadiene rubber, SBR: styrene-butadiene rubber, SAN: poly(styrene-co-acrylonitrile), NBR: nitrile-butadiene rubber.

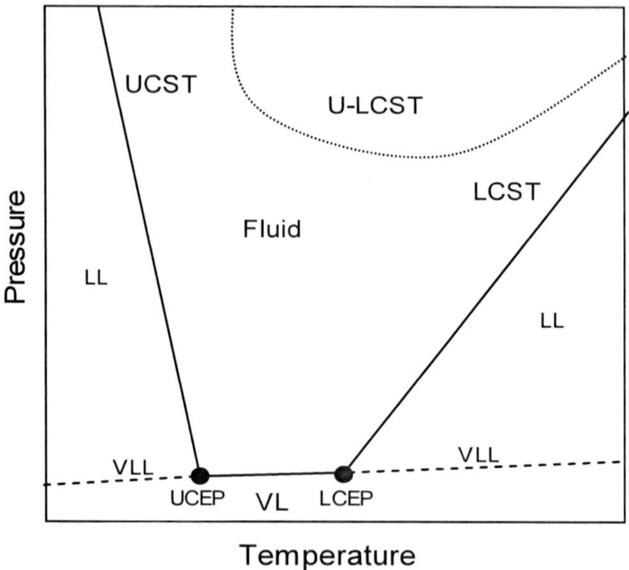

Figure 2.1. Pressure-temperature projection for the high-pressure phase behavior of polymer + solvent systems.

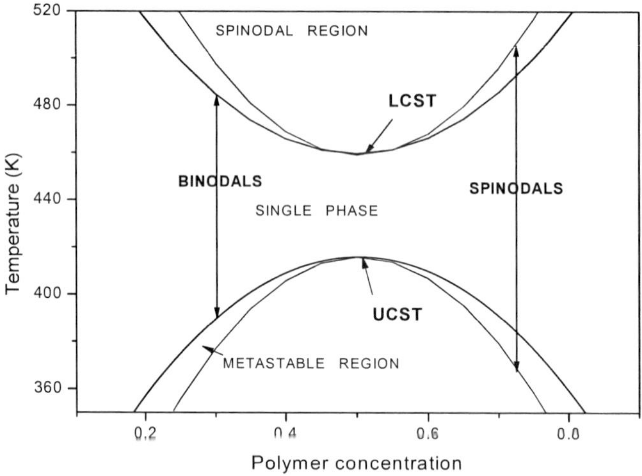

Figure 2.2. Generic phase diagram for liquid mixtures with upper and lower critical solution temperature, UCST and LCST, respectively.

Chapter 3

THERMODYNAMIC MODELS

In this chapter, we describe two non-cubic (PC-SAFT and SL) and one cubic (PR) EoS for modeling the high-pressure phase equilibria of polymer + solvent and copolymer + solvent systems.

3.1. PERTURBED CHAIN - STATISTICAL ASSOCIATING FLUID THEORY, PC-SAFT

Over the past decade, the Statistical Associating Fluid Theory (SAFT) (Huang and Radosz, 1990, 1991) has been successfully employed in the calculation of phase equilibrium of systems containing macromolecules, especially those with polymers (Chen et al., 1992; Banaszak et al., 1996; Shukla and Chapman, 1997; Pan and Radoz, 1998; Kinzl et al., 2000; Ndiaye et al., 2001; Spyriouni and Economou, 2005). The success of SAFT-EoS is due to the more rigorous modeling of molecules, which are considered as a collection of spherical segments with repulsive (hard-sphere) and attractive (dispersion) force fields. Besides, these spheres can be bound by covalent bonds to form chains (chain effect), and through specific interactions like hydrogen bonds to form short-live clusters (association effect). A detailed description of SAFT-EoS can be found elsewhere (Huang and Radosz, 1990, 1991).

A very recent version of SAFT that has appeared is that due to Gross and Sadowski (2000, 2001). Most of the terms in PC-SAFT are the same as those in the Huang and Radosz version. The term that is different is the dispersion term. However, it is not simply a different way of expressing the dispersion attraction between segments, but rather it tries to account for dispersion attraction between whole chains. Instead of adding the dispersion to hard-spheres and then forming chains, it is better to form the hard-sphere chains and then add a chain dispersion term. To do this it is required interchain rather than intersegment radial distribution functions. These are given by O'Lenick et al. (1995). The Helmholtz energy for the dispersion term is given as the sum of a first order and second order term:

$$\tilde{a} = \tilde{a}_1 + \tilde{a}_2 \tag{1}$$

where

$$\tilde{a}_1 = -2\pi\rho m^2\left(\frac{\varepsilon}{kT}\right)\sigma^3 \int_1^\infty \tilde{u}(x) g^{hc}(m; x\sigma / d) x^2 dx \tag{2}$$

$$\tilde{a}_2 = -\pi\rho m\left(1 + Z^{hc} + \rho\frac{\partial Z^{hc}}{\partial\rho}\right)^{-1} m^2\left(\frac{\varepsilon}{kT}\right)^2 \sigma^3 \frac{\partial}{\partial\rho}\left[\rho\int_1^\infty \tilde{u}(x)^2 g^{hc}(m, x\sigma / d) x^2 dx\right] \tag{3}$$

where $x = r/\sigma$ and $\tilde{u}(x) = u(x)/\varepsilon$ is the reduced intermolecular potential. The radial distribution function, g^{hc} is now an interchain function rather than a segment function as before. This is a key point in PC-SAFT. The term involving compressibilities is given by

$$\left(1 + Z^{hc} + \rho\frac{\partial Z^{hc}}{\partial\rho}\right) = \left[1 + m\frac{8\eta - 2\eta^2}{(1-\eta)^4} + (1-m)\frac{20\eta - 27\eta^2 + 12\eta^3 - 2\eta^4}{\left((1-\eta)(2-\eta)\right)^2}\right] \tag{4}$$

We still need to solve the integrals in Eqs. (2) and (3). Setting

$$I_1 = \int_1^\infty \tilde{u}(x) g^{hc}(m; x\sigma / d) x^2 dx \tag{5}$$

$$I_2 = \frac{\partial}{\partial \rho}\left[\rho \int_1^{\infty} \tilde{u}(x)^2 g^{hc}(m; x\sigma / d) x^2 dx\right] \tag{6}$$

By substituting the Lennard-Jones potential and the radial distribution function of O'Lenick et al. (1995) and doing for the series of *n*-alkanes, the integrals were fit as a power series:

$$I_1 = \sum_{i=0}^{6} a_i \eta^i \tag{7}$$

$$I_2 = \sum_{i=0}^{6} b_i \eta^i \tag{8}$$

with

$$a_i = a_{oi} + \frac{m-1}{m} a_{1i} + \frac{m-1}{m}\frac{m-2}{m} a_{2i} \tag{9}$$

$$b_i = b_{oi} + \frac{m-1}{m} b_{1i} + \frac{m-1}{m}\frac{m-2}{m} b_{2i} \tag{10}$$

Eqs. (9) and (10) require a total of 42 constants which are adjusted to fit experimental pure component data of *n*-alkanes. This direct fitting to experimental data to some extent accounts for errors in the reference equation of state, the perturbing potential and the radial distribution function which appear in the integrals of Eqs. (5) and (6). The dispersion potential given by Eqs. (2) and (3) is readily extended to mixtures using the van der Waals one-fluid theory.

For mixtures, PC-SAFT EoS (Gross and Sadowski, 2000, 2001) has the reference hard-sphere chain and the perturbation contribution terms,

$$\tilde{a}^{res} = \tilde{a}^{hc} + \tilde{a}^{pert} \tag{11}$$

where $\tilde{a} = A / NkT$. The hard chain contribution (Chapman et al., 1990) was based in the first-order thermodynamic perturbation theory:

$$\tilde{a}^{hc} = \overline{m}.\tilde{a}^{hs} - \sum_i x_i (m_i - 1). \ln g_{ii}^{hs}(\sigma_{ii}) + \tilde{a}^{ideal} \tag{12}$$

where m, x and g^{hs} are the segment number, mole fraction and the radial pair distribution function, respectively, and $\overline{m}$ is the arithmetic average of the segment number.

The contribution of hard-sphere, $\tilde{a}^{hs}$, depends on the temperature-dependent segment number, d, and the total number density of molecules, ρ, where d_i is calculated as $d_i = \sigma_{ii}[1 - 0.12.\exp(-3\varepsilon_{ii} / kT)]$ and the perturbation contribution (Barker and Henderson, 1967) is predicted from the first ($\tilde{a}_1$) and second-order ($\tilde{a}_2$) perturbation terms (Eq. 1), where $\tilde{a}_1$ and $\tilde{a}_2$ depend on the total number density of molecules, ρ, the average of the segment number, $\overline{m}$, the reduced density, η, and the van der Waals one-fluid mixing rules given by

$$\overline{m_{i,j}\varepsilon_{i,j}^l \sigma_{i,j}^k} = \sum_i \sum_j x_i x_j m_i m_j \left(\frac{\varepsilon_{ij}}{kT} \right)^l \sigma_{ij}^k \tag{13}$$

Conventional combining rules are used to determine the cross parameters:

$$\sigma_{ij} = (1/2)(\sigma_{ii} + \sigma_{jj}) \tag{14}$$

$$\varepsilon_{ij} = \sqrt{\varepsilon_{ii}\varepsilon_{jj}}\,(1 - \kappa_{ij}) \tag{15}$$

where κ_{ij} is an adjustable parameter used to evaluate the segment-segment interactions. So, m, σ and ε are the pure-component parameters for the PC-SAFT model.

Since this first PC-SAFT paper appeared, a series of further papers has appeared, applying PC-SAFT to polymers (Gross and Sadowski, 2002, Tumakaka et al., 2002), associating fluids (Gross and Sadowski, 2001a) and copolymers (Gross et al., 2003).

For copolymers, PC-SAFT EoS admits that the regular components and the homopolymers form part of chains of spherical segments of the same type. This molecular model is spread out to copolymers (Gross et al., 2003) allowing only different types of segments (segments type α and β) in molecular chain, such it is represented in Figure 3.1.

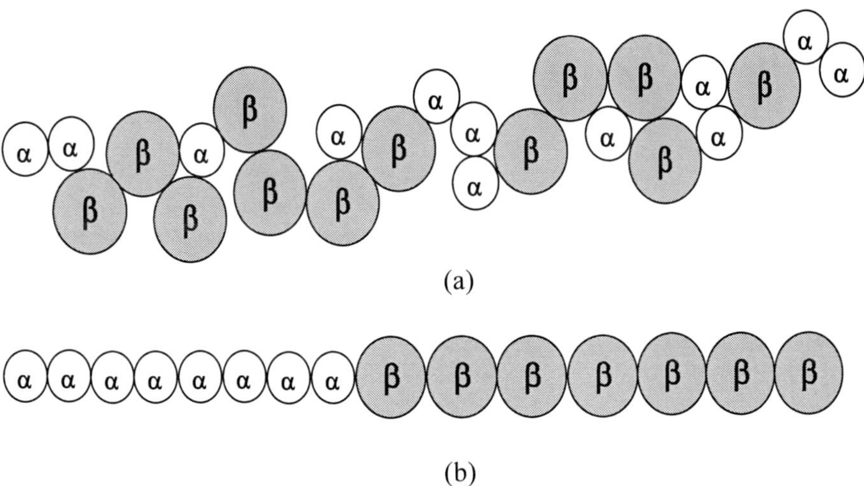

(a)

(b)

Figure 3.1. Molecular model for a random copolymer (a) type poly(α -co-β) and for alternating copolymer (b) of type poly(α-b-β), containing segments α and β.

For copolymers, the Helmholtz residual energy consists of two terms: the reference contribuition of hard chains and the dispersion contribution of segments

$$\tilde{a}^{res} = \tilde{a}^{hc} + \tilde{a}^{disp} \tag{16}$$

The segment number $m_{i\alpha}$ of segment type α is obtained from pure-component parameter $(m/MW)_{i\alpha}$,

$$m_{i,\alpha} = \frac{w_{i,\alpha} MW_{copolym}}{(m/MW)_{i,\alpha}} \tag{17}$$

where $MW_{copolym}$ is the total molecular weight of copolymer, $w_{i\alpha}$ is the mass fraction of monomers type α, and $(m/MW)_{i\alpha}$ is the segment number of type α per mass of monomer α, respectively. The total number of molecular segments, m_i, of copolymer i is the addition of all segments, according to

$$m_i = \sum_{\alpha} m_{i,\alpha} \tag{18}$$

The segment fraction, $z_{i\alpha}$, is given by

$$z_{i,\alpha} = \frac{m_{i,\alpha}}{m_i} \tag{19}$$

The polymers studied in this work are copolymers whose monomers have a random arrangement with a statistical distribution of repetitive units a and b. For thermodynamic modeling, the following suppositions are done: (1) if the segment fraction β along the main chain is lesser than the segment fraction of type α, it is assumed that all segments β are joined to segments α and that the contacts β-β are not taken into account; and (2) it is assumed that copolymer molecules with similar number of segments α and β present a strict alternative sequence of segments α and β. Summarizing, for the specific case of a copolymer with a statistical distribution of repeated units, the bonding fractions $B_{i\alpha,j\beta}$ can be determined from the next table (Gross et al., 2003).

Table 3.1. Bonding fractions, $B_{i\alpha,i\beta}$, for copolymer *i* containig segments α and β

Copolymer	Repeated unit composition	$B_{i\alpha,i\beta}$	$B_{i\alpha,i\alpha}$	$B_{i\beta,i\beta}$
Random	$z_{i,\beta} < z_{i,\alpha}$	$2[(z_{i,\beta}\, m_i)/(m_i - 1)]$	$1 - B_{i\alpha,i\beta} - B_{i\beta,i\alpha}$	0
Random	$z_{i,\beta} > z_{i,\alpha}$	$2[(z_{i,\alpha}\, m_i)/(m_i - 1)]$	0	$1 - B_{i\alpha,i\beta} - B_{i\beta,i\alpha}$
Alternating	$z_{i,\beta} = z_{i,\alpha}$	1	0	0

The reference term of hard chain is

$$\tilde{a}^{hc} = \overline{m}\tilde{a}^{hs} - \sum_i x_i (m_i - 1) \sum_{\alpha} \sum_{\beta} B_{i\alpha,i\beta} . \ln g^{hs}_{i\alpha,i\beta}(d_{i\alpha,i\beta}) \tag{20}$$

where the hard-sphere Helmholtz energy is given by

$$\tilde{a}^{hs} = \frac{1}{\xi_0}\left[\frac{3\xi_1\xi_2}{(1-\xi_3)} + \frac{\xi_2^3}{\xi_3(1-\xi_3)^2} + \left(\frac{\xi_2^3}{\xi_3^2} - \xi_0\right)\ln(1-\xi_3)\right] \quad (21)$$

and where $\overline{m}$ is the average segment number in the mixture

$$\overline{m} = \sum_i x_i m_i \sum_\alpha z_{i,\alpha} = \sum_i x_i m_i \quad (22)$$

whose radial distribution function is

$$g_{i\alpha,j\beta}^{hs}(d_{i\alpha,j\beta}) = \frac{1}{(1-\xi_3)} + \left(\frac{d_{i\alpha}d_{j\beta}}{d_{i\alpha}+d_{j\beta}}\right)\frac{3\xi_2}{(1-\xi_3)^2} + \left(\frac{d_{i\alpha}d_{j\beta}}{d_{i\alpha}+d_{j\beta}}\right)^2 \frac{2\xi_2^2}{(1-\xi_3)^3} \quad (23)$$

and where

$$\xi_k = \frac{\pi}{6}\rho\sum_i x_i m_i \sum_\alpha z_{i\alpha} d_{i\alpha}^k \quad , \quad k = (0,1,2,3) \quad (24)$$

The temperature-dependent segment diameter, $d_{i\alpha}$, of segment type α, is given by

$$d_{i\alpha} = \sigma_{i\alpha}\left[1 - 0.12.\exp(-3\varepsilon_{i\alpha}/kT)\right] \quad (25)$$

The dispersion term was given by eq. (1) and the auxiliary variables $\tilde{a}_1$ and $\tilde{a}_2$ are defined as

$$\tilde{a}_1 = -2\pi\rho.I_1(\eta,\overline{m}).\sum_i\sum_j x_i x_j m_i m_j \sum_\alpha\sum_\beta z_{i\alpha} z_{j\beta}\left(\frac{\varepsilon_{i\alpha,j\beta}}{kT}\right)\sigma_{i\alpha,j\beta}^3 \quad (26)$$

$$\tilde{a}_2 = -\pi\rho\bar{m}\left(1+Z^{hc}+\rho\frac{\partial Z^{hc}}{\partial\rho}\right)^{-1} I_2(\eta,\bar{m})\sum_i\sum_j x_i x_j m_i m_j \sum_\alpha\sum_\beta z_{i\alpha} z_{j\beta}\left(\frac{\varepsilon_{i\alpha,j\beta}}{kT}\right)^2 \sigma^3_{i\alpha,j\beta} \tag{27}$$

A non-associated copolymer (index i) requires pure-component parameters of all segment types (Gross et al., 2003); for example, the segment diameter, $\sigma_{i,\alpha}$, the segment number, $m_{i,\alpha}$ of type α in the chain and the energy parameter, $\varepsilon_{i,\alpha}/k$ (index corresponds to all constituents of copolymer). Since copolymers contain segments of different types, an analogous rule is needed, similar to the mixing rules, for a pure copolymer. Mixing rules of van der Waals one-type fluid are adopted for the dispersion term:

$$\varepsilon_{i\alpha,i\beta} = \sqrt{\varepsilon_{i\alpha}\varepsilon_{i\beta}}\left(1-\kappa_{i\alpha,i\beta}\right) \tag{28}$$

$$\sigma_{i\alpha,i\beta} = \frac{1}{2}\left(\sigma_{i\alpha}+\alpha_{i\beta}\right) \tag{29}$$

An internal parameter, $\kappa_{i\alpha,j\beta}$, is used to correct the crossing energy by the dispersion between the different segment types.

The one-fluid mixture concept is applied to the second-order compressibility term for the perturbation term; i.e.

$$\left(1+Z^{hc}+\rho\frac{\partial Z^{hc}}{\partial\rho}\right) = \left[1+\bar{m}\frac{8\eta-2\eta^2}{(1-\eta)^4}+(1-\bar{m})\frac{20\eta-27\eta^2+12\eta^3-2\eta^4}{[(1-\eta)(2-\eta)]^2}\right] \tag{30}$$

In these equations, the integrals are substituted on the pair radial distribution function of molecule chain through the sixth-order power series in density,

$$I_1(\eta,\bar{m}) = \sum_{k=0}^{6} a_k(\bar{m}).\eta^k \tag{31}$$

$$I_2(\eta,\overline{m}) = \sum_{k=0}^{6} b_k(\overline{m}).\eta^k \tag{32}$$

where $a_k(\overline{m})$ and $b_k(\overline{m})$ are the coefficients of a power series in density, and each one depends on the segment number. Liu and Hu (1996) proposed a relation to describe the dependence of each coefficient as power series in segment numbers, such as

$$a_k(\overline{m}) = a_{0k} + \frac{\overline{m}-1}{\overline{m}} a_{1k} + \frac{\overline{m}-1}{\overline{m}} \frac{\overline{m}-2}{\overline{m}} a_{2k} \tag{33}$$

$$b_k(\overline{m}) = b_{0k} + \frac{\overline{m}-1}{\overline{m}} b_{1k} + \frac{\overline{m}-1}{\overline{m}} \frac{\overline{m}-2}{\overline{m}} b_{2k} \tag{34}$$

These constants form the model; a_{0k}, a_{1k}, and a_{2k}, as well as b_{0k}, b_{1k}, and b_{2k} were fitted with thermophysical properties of pure n-alkanes (Gross and Sadowski, 2000).

3.2. Sanchez-Lacombe, SL.

The Sánchez-Lacombe (SL) lattice-gas EoS (Sánchez and Lacombe, 1976, 1978) is composed of a van der Waals-type attractive term and a lattice-gas repulsive term, and can be written in reduced form as:

$$\rho_R^2 + P_R + T_R\left[\ln(1-\rho_R) + \left(1-\frac{1}{r}\right)\rho_R\right] = 0 \tag{35}$$

where P_R, T_R and ρ_R are the reduced pressure, temperature and density, respectively, defined as $P_R = P/P^*$, $T_R = T/T^*$ and $\rho_R = \rho/\rho^*$, where T^*, P^* and ρ^* are the three pure characteristic parameters defined as $T^* = \varepsilon^*/R$, $P^* = \varepsilon^*/v^*$ and $\rho^* = MW.P^*/(RT^*r)$, where r is the number of lattice sites occupied by a molecule.

For mixtures, it is necessary to define a characteristic mixture temperature, pressure and close-packed molar volume. The characteristic mixture temperature is:

$$T^*_{mix} = \frac{\varepsilon^*_{mix}}{R} \tag{36}$$

where the mixing rule for ε_{mix}* is:

$$\varepsilon^*_{mix} = \frac{1}{v^*_{mix}} \sum_i \sum_j \varphi_i \varphi_j \varepsilon^*_{ij} v^*_{ij} \tag{37}$$

and the cross term, ε_{ij}*, is:

$$\varepsilon^*_{ij} = \sqrt{\varepsilon_{ii}\varepsilon_{jj}}(1-\kappa_{ij}) \tag{38}$$

where κ_{ij} is an adjustable parameter which takes account interactions between molecule *i* and *j*. The volume fraction of component *i*, φ_i, is:

$$\varphi_i = \frac{w_i / \rho_i v_i}{\sum_j (w_j / \rho_j v_j)} \tag{39}$$

Density can be expressed as:

$$v_R = \frac{1}{\rho_R} = \frac{V}{V^*} \tag{40}$$

where $V^* = N(rv^*_{mix})$. The mixing rule for v_{mix}* is:

$$v^*_{mix} = \sum_i \sum_j \varphi_i \varphi_j v^*_{ij} \tag{41}$$

where the cross term, $v_{ij}*$, is the arithmetic mean of the two pure-component characteristic volumes:

$$v_{ij}^{*} = (1/2)(v_{ii}^{*} + v_{jj}^{*}) \tag{42}$$

The mixing rule for the number of sites that a mixture occupies (r_{mix}) is:

$$\frac{1}{r_{mix}} = \sum_{i} \frac{\varphi_i}{r_i} \tag{43}$$

The characteristic mixture pressure is:

$$P_{mix}^{*} = \frac{RT_{mix}^{*}}{V_{mix}^{*}} \tag{44}$$

3.3. Peng-Robinson, PR

The PR EoS (Peng and Robinson, 1976) can be written as:

$$P = \frac{RT}{(\underline{V} - b)} - \frac{a}{\left[\underline{V}(\underline{V} + b) + b(\underline{V} - b)\right]} \tag{45}$$

The PR EoS parameters (a and b) are calculated with the following mixing rules:

$$a = \sum_{i} \sum_{j} x_i x_j a_{ij} \quad ; \qquad b = \sum_{i} x_i b_i \tag{46}$$

The cross term, a_{ij}, is

$$a_{ij} = \sqrt{a_i a_j}\,(1 - \kappa_{ij}) \tag{47}$$

where κ_{ij} is an adjustable parameter. The pure-component parameters (a_i and b_i) are calculated from pure-component critical properties:

$$a_i = 0.4572.a_{(T_R)}.\left(R.T_{C,i}\right)^2 / P_{C,i} \quad ; \qquad b_i = 0.0778 R.T_{C,i} / P_{C,i} \tag{48}$$

where $.a_{(T_R)} = \left[1 + m\left(1 - \sqrt{T_R}\right)\right]^2$ and parameter m is defined in terms of the acentric factor:

$$m = 0.3746 + 1.5423\,w - 0.2699\,w^2 \tag{49}$$

In this study, since copolymers do not have critical properties, the evaluation of the energy and co-volume parameters of the pure polymer and copolymer, a and b, respectively (Eq. 48) in the PR EoS (Eq. 45) is obtained by fitting the available liquid pressure-volumen-temperature (PVT) data with a single set of (a/MW) and (b/MW) parameters for several MWs of polymer or copolymer (Louli and Tassios, 2000).

Chapter 4

PHASE EQUILIBRIUM AT HIGH-PRESSURES

At high-pressures, it is necessary to apply the phi-phi approach. So, in solubility predictions, it was assumed that polymer (PS or moltem polymer) is a monodisperse polymer and that it does not dissolve in the gas phase. In this case, the conventional fugacity relations for solubility calculations with EoS can be applied.

$$x_i^L \hat{\phi}_i^L{}_{(T,P,x_i^L)} = \phi^{SCF}{}_{(T,P)} \tag{50}$$

where x is the mole fraction of component i in liquid phase and ϕ^{SCF} is the fugacity coefficient of supercritical fluid.

For the case of biodegradable polymers and copolymers, the criterium for the liquid-fluid equilibrium (LFE) is applied; with the EoS model defined as above for mixtures, the conventional fugacity relations for LFE calculations with EoS can be applied as follows

$$x_i^L \hat{\phi}_i^L{}_{(T,P,x_i^L)} = x_i^F \hat{\phi}_i^F{}_{(T,P,x_i^F)} \tag{51}$$

where x is composition, L and F represent the liquid and fluid phase, respectively. In both cases, the partial fugacity coefficient, $\hat{\phi}_i$, required for phase equilibrium calculations (Eqs. 50 and 51), is calculated from the exact thermodynamic relationship

$$ln\hat{\phi}_i = \frac{1}{RT}\int_v^{\infty}\left[\left(\frac{\partial P}{\partial n_i}\right)_{T,V,n_{i\neq j}} - \frac{RT}{v}\right]dv - ln Z \quad (52)$$

In this work, the fugacity coefficient of component i is obtained by numerical differentiation of the pressure with respect to the mole number of the respective component and then integrating Eq. (52). This numerical differentiation $[(\partial P/\partial n_i)_{T,V,ni\neq j}]$ allows the combination of any EoS with any mixing rule and eliminates the need for the cumbersome analytical determination of this property. If the numerical derivative is computed, it has to be kept in mind that P is a function of temperature, molar volume and mole fractions, whereas the partial derivative of P with respect to the mole number of component i at temperature, total volume and mole numbers of all components constant except mole number of component i. Therefore, the mole fractions x_i are converted to mole numbers n_i by multiplication with an arbitrary total mole number of $n_T = 1.00$ mole (Stockfleth and Dohrn, 1998).

$$n_i = n_T x_i \quad (53)$$

In order to obtain a derivative with respect to n_i, the mole number of component i is perturbed; the operturbation id made by a finite increment of $\Delta n_i = 0.000001$ moles, while the mole number of all other components remain constant.

$$n_i^{'} = n_i + \Delta n_i \quad (54)$$

This yields new mole fractions:

$$x_i^{'} = \frac{n_i^{'}}{n_T + \Delta n_i} \quad (55)$$

Moreover, it has to be taken into account that the partial derivative of P is taken at constant total volume. If the total mole number is increased by the increment Δn_i at constant total volume V, then the molar volume, v decreases by:

$$V = v' n_T' = v n_T \Leftrightarrow v' = v \frac{n_T}{n_T'} = v \frac{n_T}{n_T + \Delta n_i} \quad (56)$$

Now, the partial derivative can be approximated numerically, where x is the vector of mole fractions,

$$\left(\frac{\partial P}{\partial n_i} \right)_{T,V,n_{i \neq j}} = \frac{P(T, v', x') - P(T, v, x)}{\Delta n_i} \quad (57)$$

This simple two-point differentiation scheme can be expanded to a four-point scheme with a global error proportional to Δn_i using the abbreviation defined in Eq. (59),

$$\left(\frac{\partial P}{\partial n_i} \right)_{T,V,n_{i \neq j}} = \frac{P(-2\Delta n_i) - 8P(-1\Delta n_i) + 8P(+1\Delta n_i) - P(+2\Delta n_i)}{12\Delta n_i} \quad (58)$$

$$P(k\Delta n_i) = P\left(T, v' = \frac{v}{1 + k.\Delta n_i}, x' = \frac{n'}{1 + k.\Delta n_i} \right) \quad (59)$$

In Eq. (57), P is calculated using an EoS with a mixing rule by substituting numerical values of terms T, v' and x'. Eq. (58) was shown to be more stable and exact that traditional differentiations (central, forward and backward) when $\Delta n_i = 0.000001$ moles.

For polymer blends, the coexistence curve in liquid-liquid equilibria (LLE) is found from the following conditions

$$\Delta \mu_i^{L_I} = \Delta \mu_i^{L_{II}} \quad (60)$$

where $\Delta\mu$ is the change in chemical potential upon isothermally transferring component i from the pure state to the mixture, L_I and L_{II} denote the two liquid phases at equilibrium and the chemical potential can be calculated from the thermodynamic relationship.

$$\mu_i = \left(\frac{\partial G}{\partial n_i} \right)_{T,P,n_{j \neq ii}} \tag{61}$$

where n_i is the mole number of component i and G is the Gibbs free energy of mixture, which is derived using the PC-SAFT and SL EoS.

Chapter 5

RESULTS AND DISCUSSIONS

5.1. EOS PURE-COMPONENT PARAMETERS

5.1.1. Low Molecular Weight Pure-Component Parameters

For volatile component, that is, low molecular weight fluids or solvents, PC-SAFT and SL pure-component parameters were obtained by fitting vapor pressure and molar volume data for saturated pure liquid (DIPPR, 2000). The minimization function used in the modified maximum likelihood (Niesen and Yesavage, 1989; Stragevitch and d'Avila, 1997) was

$$\frac{1}{NP}\sum_{i}^{NP}\frac{\left|P_{\exp,i}^{sat}-P_{calc,i}^{sat}\right|}{P_{\exp,i}^{sat}} \tag{62}$$

The fitted characteristic parameters of each thermodynamic model were then used to calculate the average deviation from experimental specific volume,

$$\frac{1}{NP}\sum_{i}^{NP}\frac{\left|v_{\exp,i}^{l}-v_{calc,i}^{l}\right|}{v_{calc,i}^{l}} \tag{63}$$

where NP is the number of data points. Saturated vapor pressure and liquid volume deviations of these tables were calculated using the following criteria:

$$\Delta P^{l} = \frac{1}{NP}\sum_{i}^{NP}\frac{\left|P^{sat}_{\exp,i} - P^{satc}_{calc,i}\right|}{P^{sat}_{\exp,i}} \quad (64)$$

$$\Delta v^{l} = \frac{1}{NP}\sum_{i}^{NP}\frac{\left|v^{l}_{\exp,i} - v^{l}_{calc,i}\right|}{v^{l}_{\exp,i}} \quad (65)$$

5.1.2. High Molecular Weight Pure-Component Parameters

In the case of high molecular weight components (polymers and copolymers), since vapor pressures are not known, their parameters cannot be determined in the same way as for volatile fluids. To overcome this difficulty, Song and co-workers (Song et al., 1994, 1996) proposed a method for determining the parameters of high molecular weight components, but it is necessary to know the liquid pure-component *PVT* data of these components. These *PVT* data are fitted to obtain the EoS pure-component parameters for polymers and copolymers. But the problems appear when these data are not available in the literature. Elvassore et al. (2002) applied a group-contribution method coupled with the perturbed-hard-sphere-chain (PHSC) EoS to predict the PHSC pure-component parameters of low-molecular-weight compounds by estimating *PVT* data. They reported values of the group-contribution parameters, with which it is possible to extrapolate the PHSC pure-component parameters to high molecular weight compounds obtaining secondarily *PVT* data of that compound. The only input required for the model is the structure in terms of functional groups and the molecular weight. With the three pure-component parameters of PHSC EoS obtained for each polymer or copolymer, the *PVT* data are generated and used with the PC-SAFT, SL and PR models to predict their respective pure-component parameters over a pressure and temperature range suitable for engineering calculations. Average relative deviations in function of pressure and volume were calculated in according with Eqs. (66) and (67).

$$\Delta P^{L} = \frac{1}{NP}\sum_{i}^{NP}\frac{\left|P^{L}_{\exp,i} - P^{L}_{calc,i}\right|}{P^{L}_{\exp,i}} \quad (66)$$

$$\Delta v^L = \frac{1}{NP}\sum_i^{NP} \frac{\left| v^L_{\exp,i} - v^L_{calc,i} \right|}{v^L_{\exp,i}} \tag{67}$$

Very often, the molecular weight of a polymer or copolymer is quite disperse, but in this approximation the polydispersion is not taken into account, so a number-average molecular weight was used. According to Elvassore et al. (2002), their group contribution method is able to reproduce the experimental density in a wide range of temperatures and pressures with an average error of 5.0% at worst. In other words, their method is interesting, as only the knowledge of the molecular structure is required.

5.2. PS + Supercritical Fluid Systems

In this work, the gas solubility thermodynamic behavior of thirteen binary systems including one polymer (PS) and thirteen fluids (N_2, C_1, nC_4, iC_4, H_2, $=C_2$, CFC-11, CFC-114, CFC-12, CFC-22, HCFC-142b, HFC-134a, and HFC-152a)[1] was studied. In Table 5.2.1 are shown some physical characteristics of these binary systems.

5.2.1. Pure-Component Parameters

PC-SAFT and SL pure-component parameters for the chlorofluorocarbons, hydrochlorofluorocarbons, hydrofluorocarbons and supercritical fluids were obtained as was explained in item 5.1.1 and the PC-SAFT, SL and PR pure-component parameters for the PS were obtained in the same form as it was said in item 5.1.2. Tables 5.2.2a and 5.2.2b show a summary of the pure-component parameters and deviations in liquid PVT data obtained with the PC-SAFT, SL and PR models for the PS and the pure-component parameters and deviations in saturated molar volume and vapor pressure for chlorofluorocarbon, hydrochlorofluorocarbon, hydrofluorocarbon, and supercritical fluids, respectively.

---

[1] N_2: nitrogen, C_1: methane, nC_4: n-butane, iC_4: i-butane, H_2: hydrogen, $=C_2$: ethylene, CFC-11: trichlorofluoromethane, CFC-114: 1,2-dichloro-1,1,2,2-tetrafluoroethane, CFC-12: dichlorodifluoromethane, CFC-22: chlorodifluoromethane, HCFC-142b: 1-chloro-1,1-difluoroethane, HFC-134a: 1,1,1,2-tetrafluoroethane, HFC-152a: 1,1-difluoroethane, PS: polystyrene.

Deviations in Tables 5.2.2a and 5.2.2b were also calculated using the criteria of Eqs. (64), (65), (66) and (67). Critical parameters and acentric factor for the chlorofluorocarbon, hydrochlorofluorocarbon, hydrofluorocarbon, and supercritical fluids Table 5.2.2c) were taken from DIPPR (2000).

Table 5.2.1. Physical characteristic of PS + fluid systems

Binary system		Solubilities (g gas/g polymer)	Temperature (K)	Pressure (MPa)	Ref
PS +	N_2	0.0033 – 0.0094	313.15 – 353.15	6.45 – 18.01	[a]
	C_1	4.98 – 32.12[a]	373.35 – 461.55	4.82 – 34.82	[b,c]
	nC_4	0.0085 – 0.1130	348.15 – 473.15	0.16 – 3.01	[d]
	iC_4	0.0049 – 0.0737	348.15 – 473.15	0.14 – 2.93	[d]
	H_2	2.89 – 11.95[a]	443.20	8.11 – 31.06	[e]
	$=C_2$	5.27 – 10.30[a]	443.20	4.96 – 9.39	[e]
	CFC-11	0.003 – 2.333	373.15 – 553.15	0.07 – 8.28	[f]
	CFC-114	0.001 – 0.0101	293.15 – 353.15	0.07 – 0.72	[f]
	CFC-12	0.0014 – 1.0000	373.15 – 553.15	0.07 – 13.45	[f]
	CFC-22	0.001 – 1.500	373.15 – 553.15	0.07 – 13.78	[f]
	HCFC-142b	0.0063 – 0.1004	347.92 – 471.42	0.09 – 2.35	[g]
	HFC-134a	0.0067 – 0.0443	348.16 – 473.16	0.33 – 2.75	[g]
	HFC-152a	0.0075 – 0.0632	348.15 – 473.09	0.28 – 3.12	[g]

[a] cm^3 gas/g polymer, [a] Sato et al., 1996, [b] Lundberg et al. (1963), [c] Sada et al. (1987), [d] Sato et al. (2004), [e] Newitt and Weale (1948), [f] Gorski et al. (1986), [g] Sato et al. (2000).

Table 5.2.2a. Pure-component parameters for PS

EoS	Pure component parameters	PS
PC-SAFT	m/MW $(10^{-3}$ kg/mol$)^{-1}$	0.0332
	σ (m × 10^{10})	3.50
	ε/k (K)	320.14
	$\Delta P^{L\ a}$	0.0214
	$\Delta v^{L\ b}$	0.2845
SL	T* (K)	731.25
	P* (MPa)	361.23
	ρ* (kg/m^3)	1112.36
	$\Delta P^{L\ a}$	0.5698
	$\Delta v^{L\ b}$	1.2154
PR	a/MW × 10^{-4} (m^6.MPa/kg.mol)	1.3052
	b/MW × 10^{-6} (m^3/mol)	0.9415
	$\Delta P^{L\ a}$	0.8458
	$\Delta v^{L\ b}$	1.2548

[a] $\Delta P^L = \frac{1}{NP}\sum_i^{NP}\frac{\left|P^L_{\exp,i} - P^L_{calc,i}\right|}{P^L_{\exp,i}}$, [b] $\Delta v^L = \frac{1}{NP}\sum_i^{NP}\frac{\left|v^L_{\exp,i} - v^L_{calc,i}\right|}{v^L_{\exp,i}}$

Table 5.2.2b. PC-SAFT and SL pure-component parameters for fluids

Fluids	PC-SAFT EoS					SL EoS				
	m/MW $(10^{-3}$ kg/mol$)^{-1}$	σ (m × 10^{10})	ε/k (K)	ΔP^{l} [a]	Δv^{l} [b]	T* (K)	P* (MPa)	ρ* (kg/m^3)	ΔP^{l} [a]	Δv^{l} [b]
N_2	0.0438	3.31	91.56	0.0038	0.3251	158.84	101.48	805.74	0.1526	0.2032
C_1	0.0724	3.51	140.92	0.0011	0.4225	222.42	249.35	504.48	0.1528	0.3545
nC_4	0.0398	3.71	223.18	0.0024	0.3016	402.85	321.58	737.25	0.1648	0.2985
iC_4	0.0387	3.75	216.42	0.0028	0.2841	396.75	289.74	720.84	0.1254	0.2884
H_2	0.4452	3.50	27.81	0.0320	0.7582	31.68	135.25	803.51	0.2145	0.3877
$=C_2$	0.0561	3.43	178.77	0.0045	0.4385	293.25	338.23	680.40	0.1541	0.2133
CFC-11	0.0194	4.23	298.41	0.0185	0.5874	443.15	436.44	1803.25	0.3685	0.5842
CFC-114	0.0268	4.32	103.28	0.0126	0.8450	393.92	322.78	1928.42	0.3252	0.6841
CFC-12	0.0108	5.02	320.36	0.0234	0.7818	361.96	408.42	1832.08	0.4125	0.4632
CFC-22	0.0146	4.78	319.41	0.0183	0.7541	354.26	448.21	1731.21	0.2841	0.4251
HCFC-142b	0.0283	4.02	230.26	0.0187	0.6821	368.25	429.48	1578.49	0.2015	0.2945
HFC-134a	0.0312	2.56	342.48	0.0215	0.9854	329.08	483.52	1795.52	0.2365	0.3152
HFC-152a	0.0176	5.14	357.40	0.0207	0.4829	358.40	507.63	1310.45	0.2563	0.3039

[a] $\Delta P^{l} = \frac{1}{NP}\sum_{i}^{NP}\frac{\left|P_{\exp,i}^{sat} - P_{calc,i}^{satc}\right|}{P_{\exp,i}^{sat}}$, [b] $\Delta v^{l} = \frac{1}{NP}\sum_{i}^{NP}\frac{\left|v_{\exp,i}^{l} - v_{calc,i}^{l}\right|}{v_{\exp,i}^{l}}$

Table 5.2.2c. Critical properties and acentric factor for fluids

Fluids	PR EoS		
	Tc (K)	*Pc* (MPa)	ω
N_2	126.20	3.400	0.0377
C_1	190.564	4.599	0.0116
nC_4	425.122	3.796	0.2002
iC_4	407.80	3.640	0.1835
H_2	33.19	1.313	-0.2160
$=C_2$	282.34	5.041	0.0863
CFC-11	471.20	4.408	0.1894
CFC-114	418.85	3.260	0.2521
CFC-12	384.95	4.125	0.1797
CFC-22	369.3	4.971	0.2193
HCFC-142b	410.29	4.041	0.2307
HFC-134a	374.18	4.056	0.3269
HFC-152a	386.44	4.520	0.2751

5.2.2. Binary Interaction Parameters.

Temperature-dependent binary interaction parameters, κ_{ij}, in Eq. (15), (38) and (47) were determined by fitting selective experimental solvent + SCF VLE data by using the modified maximum likelihood method (Niesen and Yesavage, 1989; Stragevitch and d'Avila, 1997) by minimizing the following objective function

$$OF = \sum_{i=i}^{NP} \left[\left(\frac{\left| P_i^{exp} - P_i^{calc} \right|}{P_i^{exp}} \right) + \left(\frac{\left| T_i^{exp} - T_i^{calc} \right|}{T_i^{exp}} \right) + \sum_{j=1}^{NC-1} \left| x_{i,j}^{exp} - x_{i,j}^{calc} \right| \right] \quad (68)$$

where *NC* is the component number and *NP* is the experimental data number. Table 5.2.3 shows the temperature-dependent binary interaction parameters for each PS + supercritical solvent system, which were correlated as

$$\kappa_{ij} = C_1 + C_2 / T \quad (69)$$

The impact of this temperature-dependence of κ_{ij} on the cross-segment energy parameter is less than 0.18% for the PC-SAFT EoS, 2.85% for the SL EoS, and 3.68% for the PR EoS in terms of the cloud point pressure deviations within the temperature range of experiment for PS + SCF systems.

5.2.3. Modeling Solubility Pressures of Binary Systems

5.2.3.1. PS + Supercritical Fluid Systems

The fluid phase behavior of the PS + N_2 system is shown in Figure 5.2.1, in terms of gas solubility as a function of pressure; the behavior of both curves is almost linear. The solubility increases when the pressure increases for all three thermodynamic models. From this figure, it is easy to see that the solubility of N_2 increases with the increasing of temperature, although, in general, the solubility of a gas usually decreases with the increasing of temperature for many polymer + gas systems. This is called reverse solubility and has been observed for gases with lower critical temperature such as helium, H_2, C_1, N_2, and oxygen (Peng et al., 2000). For this binary system, the agreement between the experimental and the correlated solubilities by PC-SAFT EoS is satisfactory (0.05%), while SL and PR EoS shown less accuracy (2.08 and 2.34%, respectively).

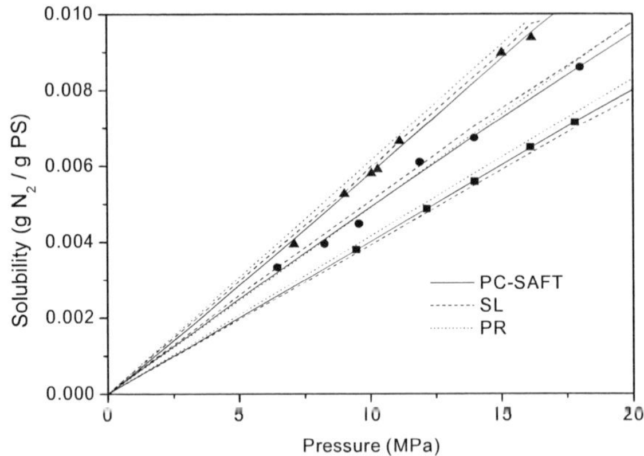

Figure 5.2.1. Solubility of N_2 in PS. Experimental data (■ = 373.20 K, ● = 413.20 K, ▲ = 453.20 K) were taken from Sato et al. (1996).

Table 5.2.3. Temperature-dependent binary interaction parameters and pressure deviations for PS + fluid systems

Binary system: PS +	EoS	κ_{ij}	ΔP^a	Binary system: PS +	EoS	κ_{ij}	ΔP^a
N_2	PC-SAFT	-0.0101 + 11.6121/T	0.05	CFC-11	PC-SAFT	0.0056 + 2.1584/T	0.13
	SL	-0.0022 – 8.5601/T	2.08		SL	0.0109 + 2.1254/T	1.85
	PR	0.0014 – 5.3044/T	2.34		PR	0.0236 - 13.416/T	2.85
C_1	PC-SAFT	0.0088 + 3.301/T	0.09	CFC-114	PC-SAFT	0.0073 + 1.8962/T	0.13
	SL	0.0095 + 4.948/T	2.36		SL	0.0075 + 1.0245/T	2.74
	PR	-0.0391 + 19.1215/T	2.07		PR	0.0402 - 10.7415/T	3.68
nC_4	PC-SAFT	0.0074 + 3.206/T	0.12	CFC-12	PC-SAFT	0.0093 + 1.2145/T	0.10
	SL	0.0083 + 4.487/T	2.68		SL	0.0093 + 3.4125/T	2.74
	PR	-0.0363 + 18.202/T	3.05		PR	0.0215 - 13.2141/T	2.23
iC_4	PC-SAFT	0.0062 + 2.735/T	0.15	CFC-22	PC-SAFT	0.0062 + 2.2252/T	0.18
	SL	0.0066 + 4.412/T	2.45		SL	0.0093 + 4.1241/T	1.96
	PR	-0.0262 + 14.152/T	3.64		PR	0.0314 - 12.4125/T	2.28

Table 5.2.3. (Continued)

Binary system: PS +	EoS	κ_{ij}	ΔP^a	Binary system: PS +	EoS	κ_{ij}	ΔP^a
$=C_2$	PC-SAFT	0.0123	0.09	HCFC-142b	PC-SAFT	0.0063 + 2.304/T	0.11
	SL	0.0281	1.86		SL	0.0093 - 1.232/T	2.03
	PR	0.0045	2.74		PR	0.0354 - 14.1363/T	2.35
H_2	PC-SAFT	0.0158	0.12	HFC-134a	PC-SAFT	0.0018 + 3.845/T	0.13
	SL	0.0362	2.85		SL	0.0068 + 4.485/T	1.56
	PR	-0.0016	3.10		PR	0.0202 - 5.1042/T	1.28
				HFC-152a	PC-SAFT	0.0071 + 2.825/T	0.09
					SL	0.0086 + 4.985/T	1.89
					PR	0.0385 - 13.1032/T	1.72

[a]
$$\Delta P = \frac{1}{NP}\sum_{i}^{NP}\frac{\left|P_{\exp,i} - P_{calc,i}\right|}{P_{\exp,i}}$$

In Figure 5.2.2 the correlated solubilities for the PS + C_1 system are shown at three temperatures. This system has the same fluid phase behavior as the PS + N_2 system. In other words, the solubility increases when pressure increases, almost in a linear variation, especially at 428.57 and 461.55 K, and the solubility increases when temperature decreases for a certain pressure. In terms of pressure deviations, PC-SAFT EoS has the lowest deviations (0.09%) against the results obtained with SL and PR EoS (2.36% and 2.07%).

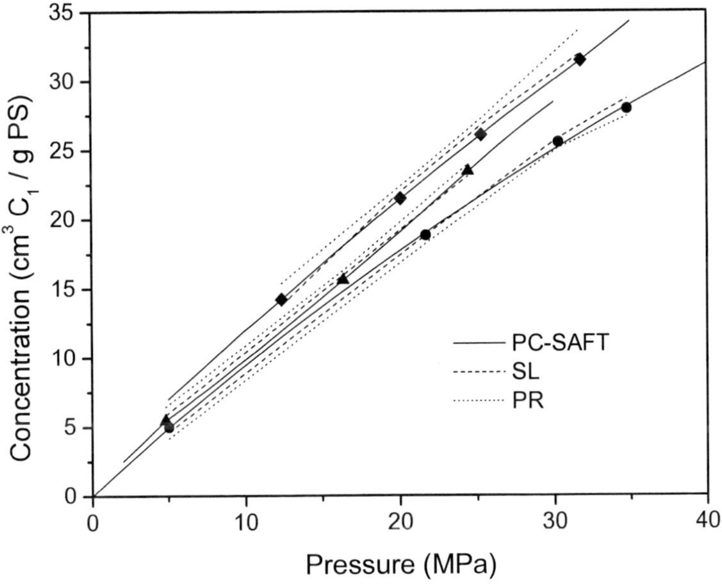

Figure 5.2.2. Solubility of C_1 in PS. Experimental data (● = 398.57 K, ▲ = 428.57 K, ◆ = 461.55 K) were taken from (Lundberg et al., 1963; Sada et al., 1987).

Solubilities of nC_4 and iC_4 in PS were modeled up to 3 MPa and along four isotherms from 348.15 to 473.15 K. For these two binary systems, the solubility increases with the increasing of pressure and the decreasing of temperature. The solubilities correlated with PC-SAFT, SL, and PR EoS, using temperature-dependent binary interaction parameters, are shown in Figures 5.2.3 and 5.2.4 for the PS + nC_4 and PS + iC_4 systems, respectively. For these systems, PC-SAFT EoS could correlate the solubility pressures with a deviation 0.15%, while SL Eos shows a deviation of 2.68%, and PR EoS shows a deviation of 3.64%.

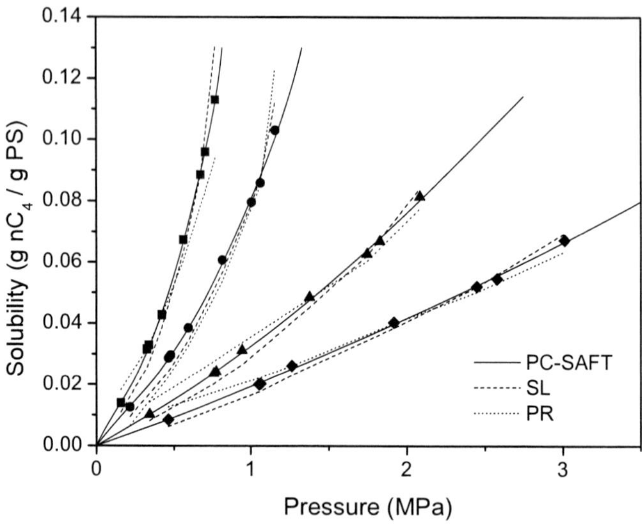

Figure 5.2.3. Solubility of nC_4 in PS. Experimental data (■ = 348.15 K, ● = 373.15 K, ▲ = 423.15 K, ◆ = 473.15 K) were taken from Sato et al. (2004).

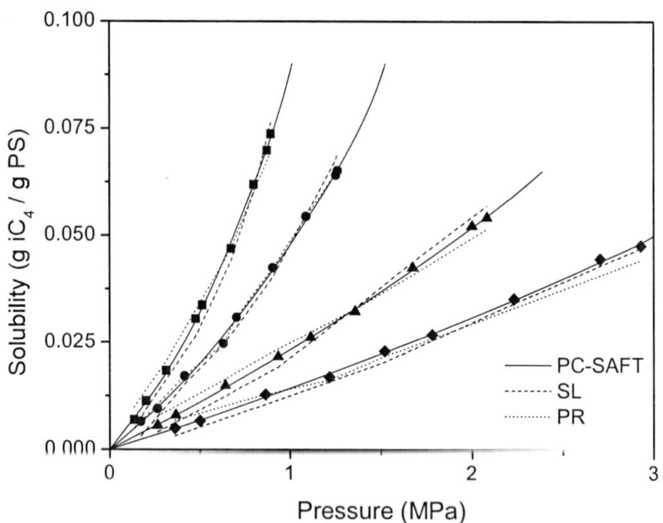

Figure 5.2.4. Solubility of iC_4 in PS. Experimental data (■ = 348.15 K, ● = 373.15 K, ▲ = 423.15 K, ◆ = 473.15 K) were taken from Sato et al. (2004).

The linear relationship between solubility and pressure of the PS + H_2 and PS + $=C_2$ systems at 443.20 K is shown in Figure 5.2.5. From this figure, it is possible to notice that the solubilities of both H_2 and $=C_2$ in PS increase when pressures increase. In terms of average relative pressure deviations, for the PS + H_2 system, PC-SAFT, SL, and PR EoS could correlate solubilities with 0.12%, 2.85%, and 3.10% deviations, respectively, while for the PS + $=C_2$ system, the deviations were 0.09%, 1.86%, and 2.74%.

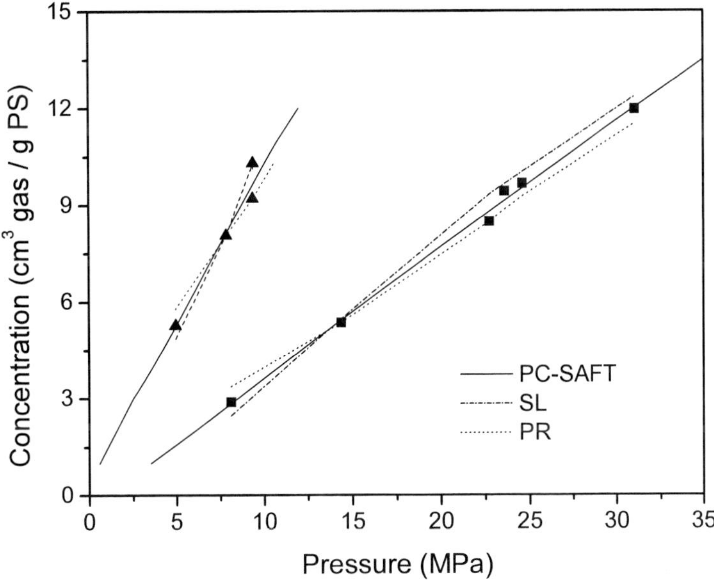

Figure 5.2.5. Solubility of H_2 in PS and $=C_2$ in PS. Experimental data (■: PS + H_2 system, ▲ : PS + $=C_2$ system) were taken from Newitt and Weale (1948).

5.2.3.2. PS + Chlorofluorocarbon Fluid Systems

The solubilities of chlorofluorocarbon fluids (CFC-11, CFC-114, CFC-12, and CFC-22) in PS at various temperatures are shown in Figures 5.2.6 to 5.2.9. For these binary systems, the relationship between solubility and pressure is again linear at lower pressures, becoming exponential at higher pressures, specifically for the CFC-11 + PS, CFC-12 + PS and CFC-22 + PS systems. The maximum average relative pressure deviations obtained by PC-SAFT, SL, and PR EoS, listed in Table 5.2.3, were 0.18%, 2.74%, and 3.68%, respectively.

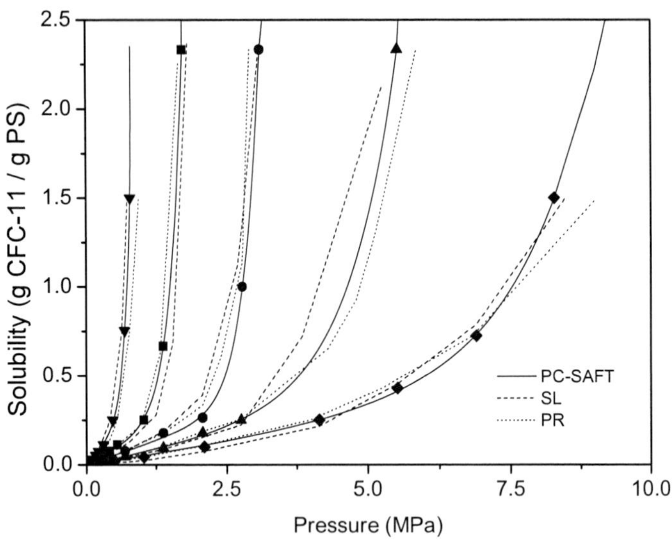

Figure 5.2.6. Solubility of CFC-11 in PS. Experimental data (▼ = 373.15 K, ■ = 413.15 K, ● = 453.15 K, ▲ = 493.15 K, ◆ = 553.15 K) were taken from Gorski et al. (1986).

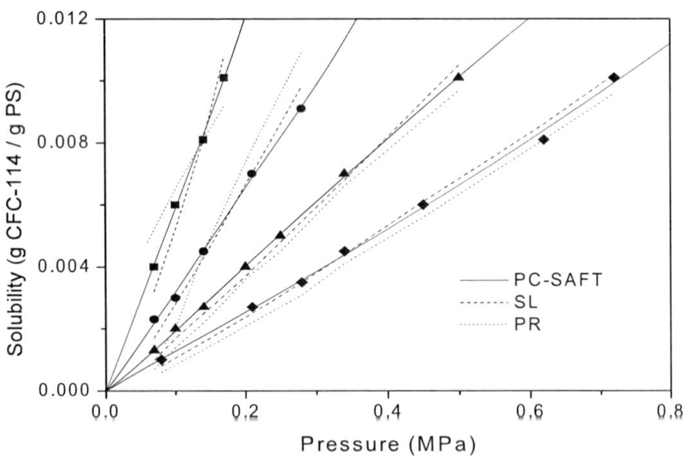

Figure 5.2.7. Solubility of CFC-114 in PS. Experimental data (■ = 293.15 K, ● = 313.15 K, ▲ = 333.15 K, ◆ = 353.15 K) were taken from Gorski et al. (1986).

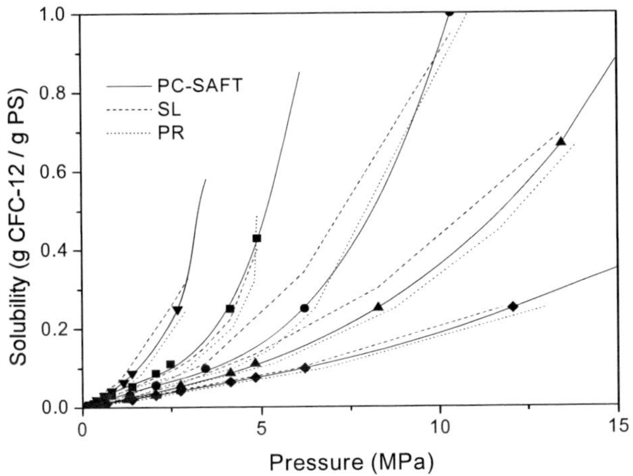

Figure 5.2.8. Solubility of CFC-12 in PS. Experimental data (▼ = 373.15 K, ■ = 413.15 K, ● = 453.15 K, ▲ = 493.15 K, ◆ = 553.15 K) were taken from Gorski et al. (1986).

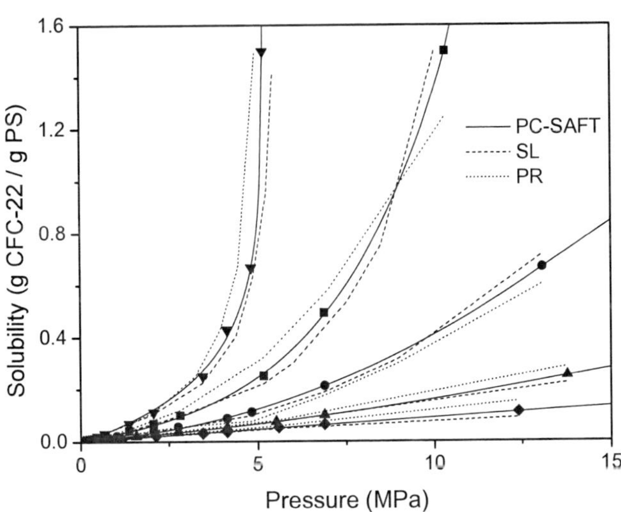

Figure 5.2.9. Solubility of CFC-22 in PS. Experimental data (▼ = 373.15 K, ■ = 413.15 K, ● = 453.15 K, ▲ = 493.15 K, ◆ = 553.15 K) were taken from Gorski et al. (1986).

5.2.3.3. PS + Hydrochlorofluorocarbon and PS + Hydrofluorocarbon Fluid Systems

Correlation results for the binary systems PS + hydrofluorocarbon (HFC-134a and HFC-152a) and hydrochlorofluorocarbon (HCFC-142b) fluids are shown in Figures 5.2.10 to 5.2.12, compared with experimental data from Sato et al. (2000). From these figures, it is easy to notice that the solubility of each blowing agent increases with pressure and decrease with temperature. As it was said previously, this behavior is exhibited for many polymer and condensable gas systems. The binary interaction parameters, κ_{ij} (Eqs. 15, 38, 47) were determined from experimental solubility pressures as function of temperature. The interaction parameters and correlation errors in terms of solubility pressure deviations for these systems are listed in Table 5.2.3. PC-SAFT, SL, and PR EoS could correlate solubilities with 0.13%, 2.03% and 2.35% average relative pressure deviations, respectively.

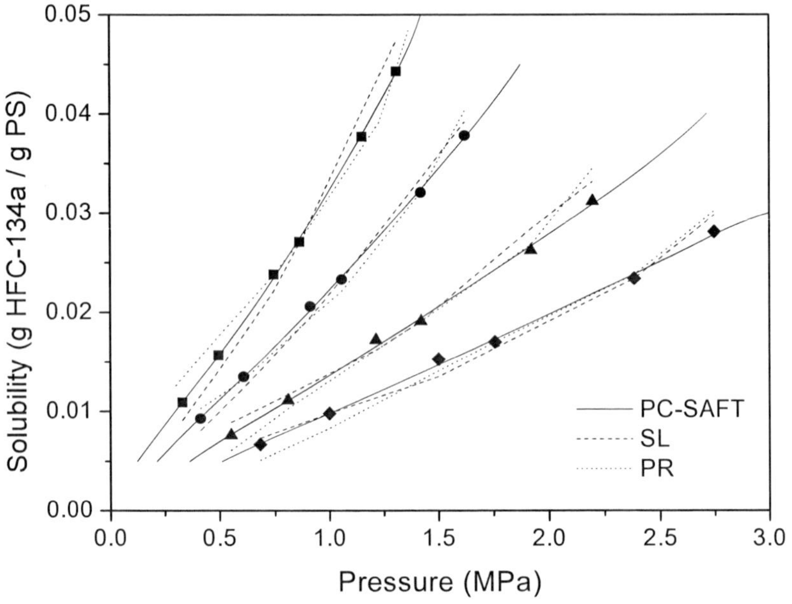

Figure 5.2.10. Solubility of HFC-134a in PS. Experimental data (■ = 348.16 K, ● = 373.16 K, ▲ = 423.15 K, ◆ = 473.16 K) were taken from Sato et al. (2000).

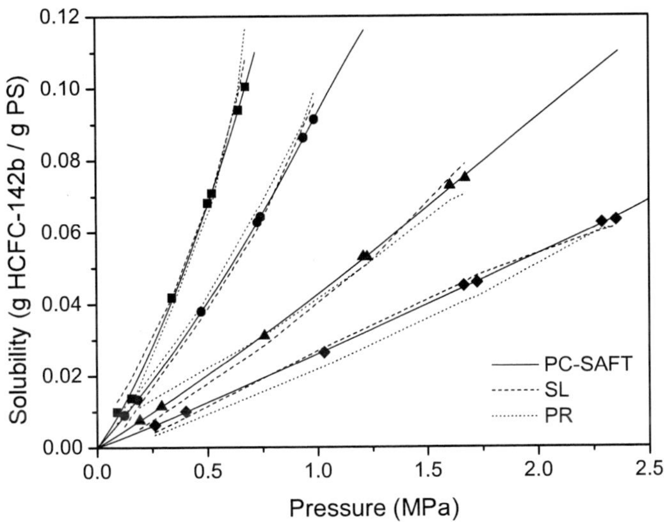

Figure 5.2.11. Solubility of HCFC-142b in PS. Experimental data (■ = 347.92 K, ● = 373.62 K, ▲= 422.15 K, ◆ = 471.42 K) were taken from Sato et al. (2000).

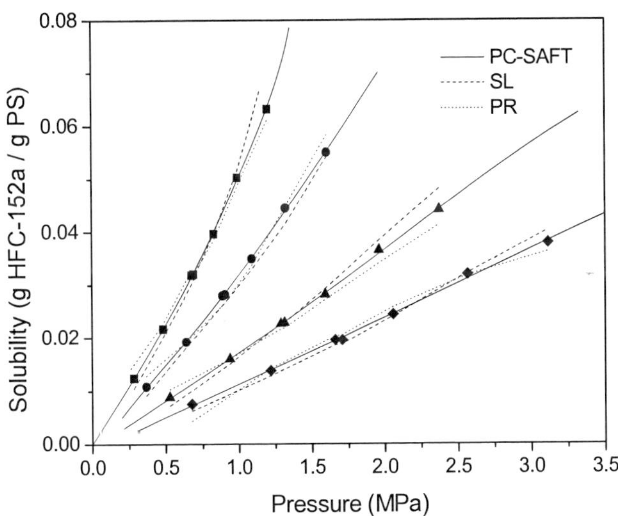

Figure 5.2.12. Solubility of HFC-152a in PS. Experimental data (■ = 348.15 K, ● = 373.13 K, ▲= 423.14 K, ◆ = 473.09 K) were taken from Sato et al. (2000).

In summary, solubilities of several gases (chlorofluorocarbon, hydrochlorofluorocarbon, hydrofluorocarbon, and supercritical fluids) in PS were modeled in a range of temperatures from 293.15 K to 553.15 K and pressures up to 35 MPa. Solubilities of all gases, except chlorofluorocarbon fluids, increase almost linearly with pressure; for chlorofluorocarbon gases, the solubility increases in an exponential way with pressure increasing. The solubilities of all gases (expect N_2) in PS show a reverse behavior due to the influence of the temperature. The solubilities were modeled by the PC-SAFT, SL and PR EoS with a temperature-dependent binary interaction parameters, κ_{ij}, and the results obtained with the PC-SAFT EoS showed a better agreement with the experimental data than the other two models.

5.3. Molten Polymer + CO_2 Systems

In this section, the modeling of solubilities of CO_2 in several molten polymers is studied. Solubility pressure calculations are carried out for eleven polymers: HPDE, LDPE, i-PP, p(VAc), PS, p(MMA), p(BMA), p(DMS), and PC[1] in CO_2 at relatively high pressures, using the PC-SAFT, and PR EoS. The gas-liquid equilibrium experimental data were taken from literature, as shown in Table 5.3.1. This table also includes the number of points, temperature and pressure ranges of each data set as well as the polymer molecular weight.

5.3.1. Evaluation of EoS Pure-Component Parameters

The pure-component parameters and the deviations from DIPPR correlation for carbon dioxide for each thermodynamic model are presented in Table 5.3.2; these deviations were calculated as described in section 5.1.1.

Parameters for pure polymers and deviations from liquid *PVT* data obtained with the PC-SAFT and PR models are presented in Tables 5.3.3a, and 5.3.3b, respectively. Liquid deviations from Tait pseudo-experimental data are also shown. PC-SAFT correlates very well the liquid specific volume, with less than 1% deviation for all polymers, and PR EoS also satisfactorily

1 HDPE: high-density polyethylene, LDPE: low-density polyethylene, i-PP: i-polypropylene, p(VAc): poly(vinyl acetate), PS: polystyrene, p(MMA): poly(methyl methacrylate), p(BMA): poly(butyl methacrylate, p(DMS): poly(dimethyl siloxane), PC: poly(carbonate bisphenol-A).

correlates it, with less than 4% deviation for all polymers. Polymer parameters were calculated as described in section 5.1.2.

Table 5.3.1. Some physical properties of CO_2 + polymer systems

Binary system: CO_2 +	NP	Temperature (K)	Pressure (MPa)	MW $_{Polymer}$	Ref
HDPE	16	293.15 - 323.15	1.88 - 4.25	100 000	[a]
	17	433.15 - 473.15	6.608 - 18.123	111 000	[b]
LDPE	5	423.15	0.655 - 3.365	250 000	[c]
i-PP	5	453.15	7.083 - 17.242	220 000	[b]
	19	433.15 - 473.15	5.419 - 17.529	451 000	[b]
p(VAc)	49	313.2 - 353.2	0.294 - 10.100	100 000	[d]
	17	313.2 - 323.2	0.898 - 8.755	100 000	[e]
	31	313.15 - 373.15	0.199 - 17.449	100 000	[f]
PS	26	373.15 - 453.15	2.472 - 20.036	187 000	[g]
	6	298.15	0.400 - 0.900	50 000	[h]
	35	373.15 - 473.15	2.068 - 20.151	330 000	[f]
p(MMA)	77	263.15 - 453.15	1.520 - 9.120	100 000	[i]
p(BMA)	63	313.2 - 353.2	0.549 - 10.200	100 000	[d]
	8	298.15	0.200 - 0.900	13 600	[h]
p(DMS)	26	308.00	0.271 - 6.282	100 000	[j]
PC	7	293.15	0.992 - 5.836	100 000	[k]
	9	313.15 - 333.15	20.00 - 40.00	64 000	[l]

[a] von Solms et al. (2004), [b] Sato et al. (1999), [c] Davis et al. (2004), [d] Wang et al. (1990), [e] Takashima et al. (1990), [f] Sato et al. (2001), [g] Sato et al. (1996), [h] Wang et al. (2003), [i] Edwards et al. (1998), [j] Pope et al. (1991), [k] Keller et al. (1999), [l] Tang et al. (2004).

Table 5.3.2. CO_2 pure-component parameters and critical properties

PC SAFT EoS				PR EoS			
m/MW (10^{-3} kg/mol)$^{-1}$	σ (10^{10} m)	ε/k (K)	$\Delta P^{l\,a}$ (%)	Tc (K)	Pc (MPa)	ω	$\Delta P^{l\,a}$ (%)
0.0482	2.7352	166.21	0.49	304.21	7.383	0.2236	1.56

$$^{a}\ \Delta P^{l} = \frac{1}{NP}\sum_{i}^{NP}\frac{\left|P_{\exp,i}^{sat} - P_{calc,i}^{satc}\right|}{P_{\exp,i}^{sat}}$$

Table 5.3.3a. Polymer pure-component parameters for PC-SAFT EoS

Polymer	m/MW $(10^{-3}$ kg/mol$)^{-1}$	σ $(10^{10}$ m)	ε/k (K)	Δv^{L} [a] (%)
HDPE	0.02819	4.0125	320.24	0.1879
LDPE	0.03391	3.7508	300.41	0.1704
i-PP	0.02453	4.2412	371.33	0.5967
p(VAc)	0.02991	3.5086	310.14	0.1298
PS	0.03324	3.5022	320.14	0.2845
p(MMA)	0.03408	3.3412	330.43	0.3542
p(BMA)	0.02616	4.0215	435.25	0.3815
p(DMS)	0.03252	3.5321	205.32	0.2812
PC	0.03719	3.1824	290.41	0.4215

[a] $\Delta v^{L} = \dfrac{1}{NP}\sum_{i}^{NP}\dfrac{\left|v^{L}_{\exp,i} - v^{L}_{calc,i}\right|}{v^{L}_{\exp,i}}$

5.3.2. Correlation of GLE Data

Table 5.3.4 presents the PC-SAFT and PR binary interaction parameters, calculated with eq. (68), as well as solubility pressure deviations obtained in the modeling. In each case examined, the average absolute deviation in solubility pressure is reported.

Table 5.3.3b. Polymer pure-component parameters for PR EoS

Polymer	a/MW $(10^{-4}$ m^{6}.MPa/kg.mol)	b/MW $(10^{-6}$ m$^{3/}$mol)	Δv^{L} [a] (%)
HDPE	1.2795	1.2046	2.9512
LDPE	1.3698	1.1842	2.5845
i-PP	1.2875	1.2386	3.4256
p(VAc)	1.8412	0.8412	1.3341
PS	1.3052	0.9415	1.2548
p(MMA)	1.2647	0.8413	1.5236
p(BMA)	1.0215	0.9315	1.5635

p(DMS)	1.0148	0.9911	1.3841
PC	1.2815	0.5318	2.0153

[a] $\Delta v^L = \frac{1}{NP}\sum_i^{NP} \frac{\left|v^L_{\exp,i} - v^L_{calc,i}\right|}{v^L_{\exp,i}}$

Table 5.3.4. Binary interaction parameters for CO_2 + polymer systems

Binary System: CO_2 +	MW polymer	PC-SAFT $\kappa_{CO_2-polym}$	PC-SAFT ΔP [a] (%)	PR $\kappa_{CO_2-polym}$	PR ΔP [a] (%)
HDPE	100 000	0.0151	2.89	0.0186	6.15
	111 000	0.0103	3.15	0.0173	7.43
LDPE	250 000	0.0084	0.85	0.0113	2.16
i-PP	220 000	-0.0012	1.25	0.0215	6.18
	451 000	0.0008	1.38	0.0153	5.26
p(VAc)	100 000	0.0035	0.89	0.0150	4.18
	100 000		1.02		3.16
	100 000		1.08		5.12
PS	187 000	0.0063	0.78	0.0193	6.18
	50 000	0.0028	0.89	0.0216	8.11
	330 000	0.0065	0.85	0.0157	7.15
p(MMA)	100 000	0.0103	0.85	0.0215	4.23
p(BMA)	100 000	0.0084	0.95	-0.0099	4.83
	13 600	0.0025	0.92	0.0218	3.05

Binary System: CO_2 +	MW polymer	PC-SAFT $\kappa_{CO_2-polym}$	PC-SAFT ΔP [a] (%)	PR $\kappa_{CO_2-polym}$	PR ΔP [a] (%)
p(DMS)	100 000	0.0153	1.06	-0.0118	7.43
PC	100 000	0.0120	1.13	0.0126	11.48
	64 000	0.0138	1.05	0.0225	9.41
Overall pressure deviations			1.18		5.56

[a] $\Delta P = \frac{1}{NP}\sum_i^{NP} \frac{\left|P_{\exp,i} - P_{calc,i}\right|}{P_{\exp,i}}$

From the results shown in Table 5.3.4, we can conclude that PC-SAFT EoS is able to calculate satisfactorily the gas-liquid behavior for these systems, since the pressure deviations vary between 0.78 to 3.15%, while the deviations obtained with the PR EoS vary between 2.16 to 11.48%.

5.3.3. HDPE + CO_2 and LDPE + CO_2 Systems

The following comments summarize our observations on the obtained results. In general, the solubility of CO_2 in molten polymers usually decreases with increasing temperatures at constant pressure for many CO_2 + polymer systems, as it is shown through Figures 5.3.1 to 5.3.9. Correlated results and experimental data (von Solms et al., 2004; Sato et al., 1999) for the solubilities of CO_2 in molten HDPE at 433.15, 453.15 and 473.15 K are shown in Figure 5.3.1. The isotherms have practically the same slope and the PC-SAFT and PR EoS are able to represent the fluid phase behavior of this binary system.

On the other hand, the fluid phase behavior of the LDPE (MW = 250 000) + CO_2 system was also modeled at 423.15 K and these results were compared with experimental data (Davis et al., 2004) as shown in Figure 5.3.2. The obtained deviations were 0.85% for PC-SAFT and 2.16% for PR.

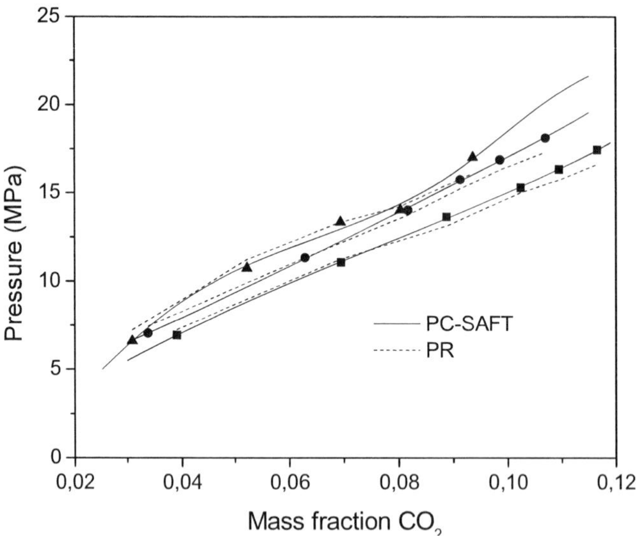

Figure 5.3.1. Solubilities of CO_2 in molten HDPE. Experimental data (■ = 433.15 K, ● = 453.15 K, ▲ = 473.15 K) were taken from Sato et al. (1999).

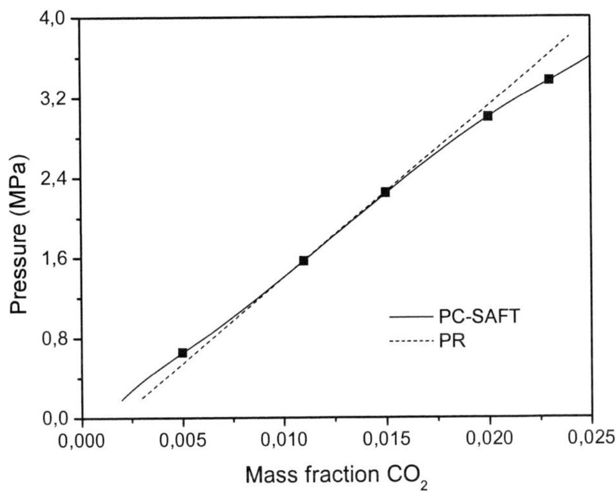

Figure 5.3.2. Solubilities of CO2 in molten LDPE. Experimental data (■ = 423.15 K) were taken from Davis et al. (2004).

5.3.4. i-PP + CO_2 Systems

Figure 5.3.3 shows the comparisons between calculated solubilities and experimental data (Sato et al., 1999) at three temperatures for the CO_2 + i-PP systems. These three isotherms have also almost the same slope and PC-SAFT is able to correlate the data with more accuracy than PR EoS in terms of pressure solubility deviations. These deviations are 1.25% (MW = 220 000) and 1.38% (MW = 451 000) for PC-SAFT, and 6.18% (MW = 220000) and 5.26% (MW = 451 000) for PR.

5.3.5. P(Vac) + CO_2 Systems

Correlated pressure solubilities and experimental data (Wang et al., 1990; Takishima et al., 1990; Sato et al., 2001) at 313.2, 323.2, 333.2 and 353.2 K for the of CO_2 + p(VAc) system are shown in Figure 5.3.4. These four isotherms join at lower CO_2 mass fractions, but at higher CO_2 mass fractions they seem to have the same slope. From this figure, it is clear that satisfactory results can be obtained with only one adjustable parameter for the PC-SAFT EoS, while the PR EoS obtains a good accuracy only at lower pressures.

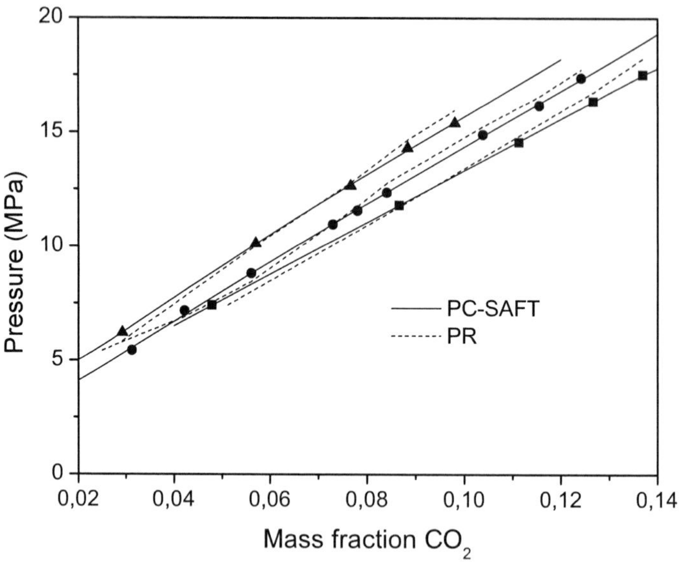

Figure 5.3.3. Solubilities of CO_2 in molten i-PP. Experimental data (■ = 433.15 K, ● = 453.15 K, ▲ = 473.15 K) were taken from Sato et al. (1999).

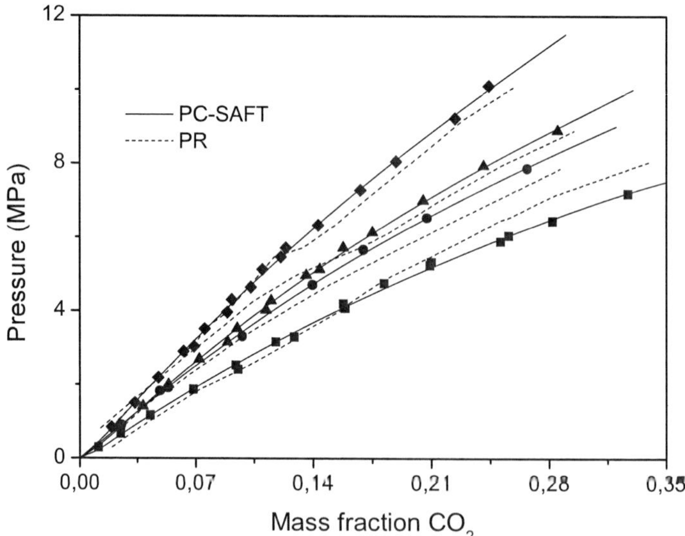

Figure 5.3.4. Solubilities of CO_2 in molten p(VAc), Experimental data (■ = 313.20 K, ● = 323.20 K, ▲ = 333.20 K, ◆ = 353.20 K) were taken from Wang et al. (1990).

5.3.6. PS + CO_2 Systems

Figure 5.3.5 shows comparisons between calculated pressure solubilities and experimental data (Sato et al., 1996; Sato et al., 2001; Wang et al., 2003) at four temperatures for CO_2 + PS system. These four isotherms have different slopes. PC-SAFT EoS was able to correlate this binary system with higher accuracy than the PR EoS (0.78% to 0.89% in pressure solubility deviations for PC-SAFT against 6.18% to 8.11% for PR).

5.3.7. p(MMA) + CO_2 Systems

Correlated pressure solubilities of CO_2 in p(MMA) and experimental data (Edwards et al., 1998) at seven temperatures are shown and compared in Figure 5.3.6. From this figure, it can be noticed that at higher temperatures, the slopes of isotherms become more pronounced and that the two models correlate this binary system with different accuracy. PR EoS is less accurate at lower temperatures, but the accuracy increases at higher temperatures. The deviation obtained with PC-SAFT EoS is 0.85%, while the deviation obtained with the PR EoS is 4.23%.

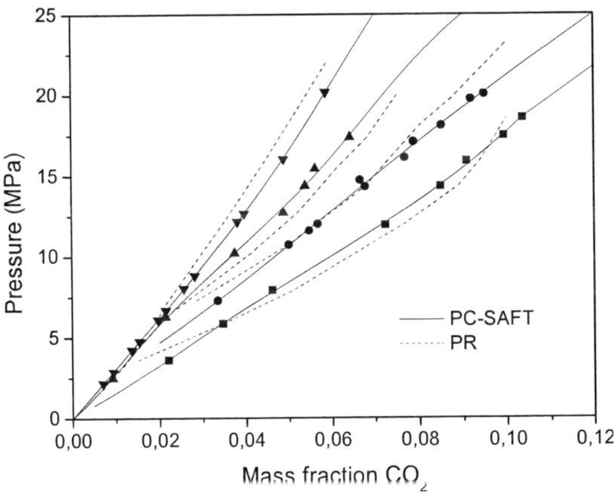

Figure 5.3.5. Solubilities of CO_2 in molten PS. Experimental data (■ = 373.15 K, ● = 413.15 K, ▲= 453.15 K, ▼ = 473.15 K) were taken from Sato et al. (1996, 2001).

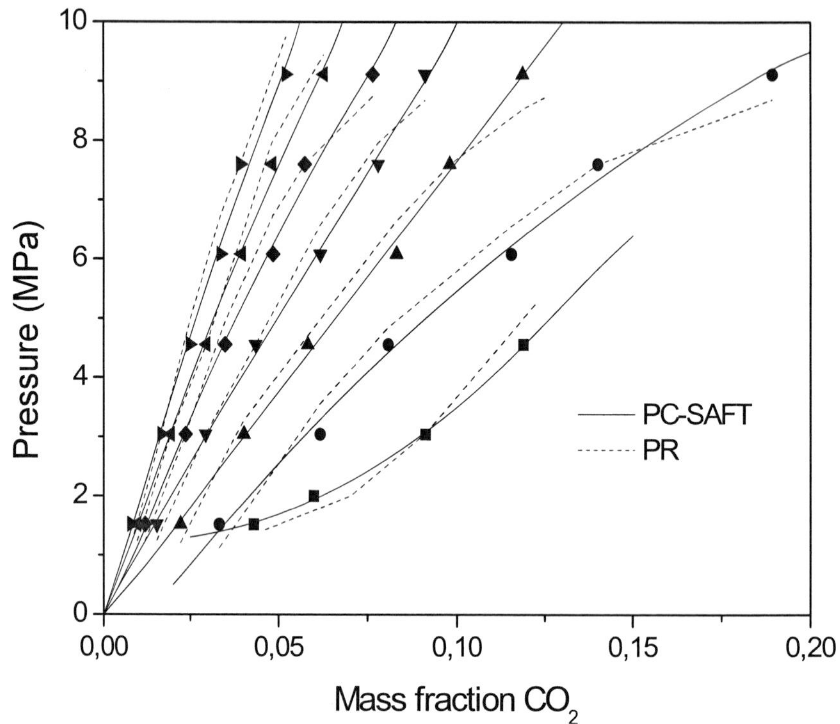

Figure 5.3.6. Solubilities of CO_2 in molten p(MMA). Experimental data (■ = 293.15 K, ● = 313.15 K, ▲ = 333.15 K, ▼ = 353.15 K, ◆ = 373.15 K, ◀ = 393.15 K, ▶ = 413.15 K) were taken from Edwards et al. (1998).

5.3.8. p(BMA) + CO_2 Systems

Figure 5.3.7 shows comparisons between calculated pressure solubilities and experimental data (Wang et al., 1990, 2003) for CO_2 + p(BMA) system at 313.2, 333.2, and 353.2 K. These three isotherms join at lower CO_2 mass fractions and have different slopes at higher CO_2 mass fractions. The relative pressure deviations obtained for the PC-SAFT EoS are between 0.92% 0.95%, while the values for the PR EoS are between 3.05% and 4.83%.

5.3.9. p(DMS) + CO_2 Systems

Figure 5.3.8 compares the calculated pressure solubilities obtained with the PC-SAFT and PR EoS with experimental data for CO_2 in p(DMS) at 308.0 K (Pope et al., 1991). From this figure, PC-SAFT has a better performance than PR. In terms of relative pressure deviations, the performance of each thermodynamic model was 1.06% and 7.43% for PC-SAFT and PR, respectively.

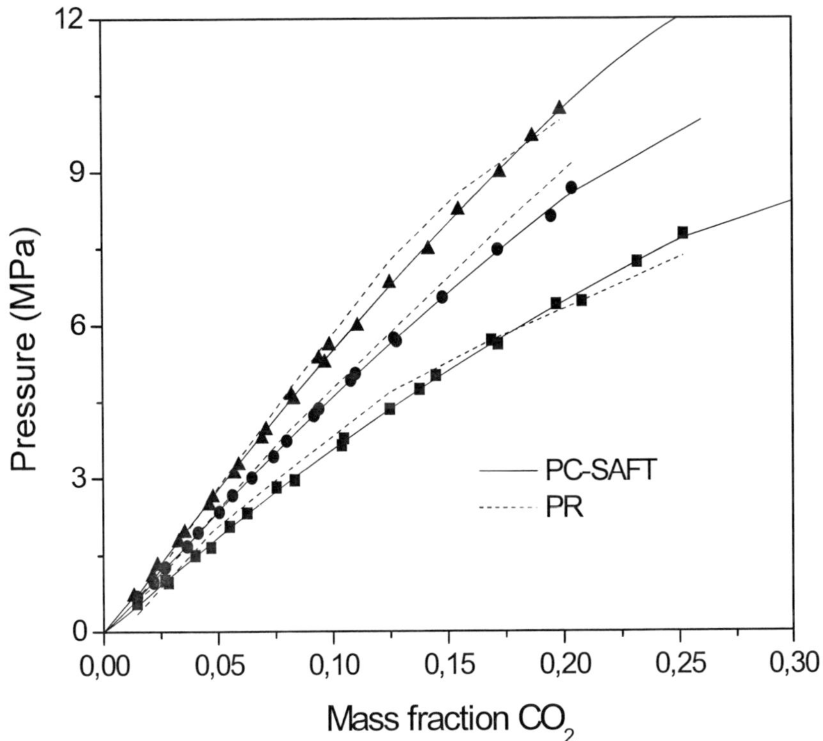

Figure 5.3.7. Solubilities of CO_2 in molten p(BMA). Experimental data (■ = 313.20 K, ● – 333.20 K, ▲ – 353.20 K) were taken from Wang et al. (1990).

5.3.10. PC + CO_2 Systems

Correlated values and experimental data of the CO_2 + PC system (Keller et al., 1999; Tang et al., 2004) at 313.15, 323.15 and 333.15 K are shown in

Figure 5.3.9. From this figure, it can be noticed that PC-SAFT is able to correlate the experimental data with higher accuracy. The relative pressure deviations obtained with the PC-SAFT and PR EoS are 1.05% to 1.13% and 9.41% to 11.48%, respectively.

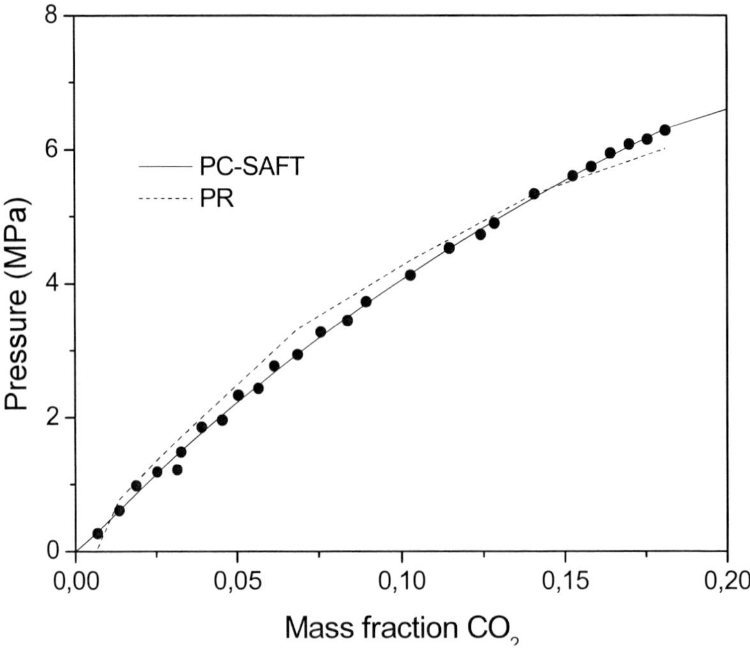

Figure 5.3.8. Solubilities of CO_2 in molten p(DMS). Experimental data (● = 308.00 K) were taken from Pope et al. (1991).

In summary, the PC-SAFT EoS gives the best overall pressure deviation results, which is not surprising for polymer systems, because this thermodynamic model regards the monomer + monomer, monomer + solvent and solvent + solvent interactions in a more rigorous form. The overall pressure deviation obtained by the PR EoS (5.56%) is higher than that obtained by the PC-SAFT EoS (1.18%), but it is important to keep in mind than the PR EoS is much simpler than PC-SAFT and requires only one interaction parameter. These results indicate that, although PR is able to correlate very well pure polymer systems, the mixture effects are not correctly predicted for highly asymmetric mixtures, such as CO_2 + polymer systems.

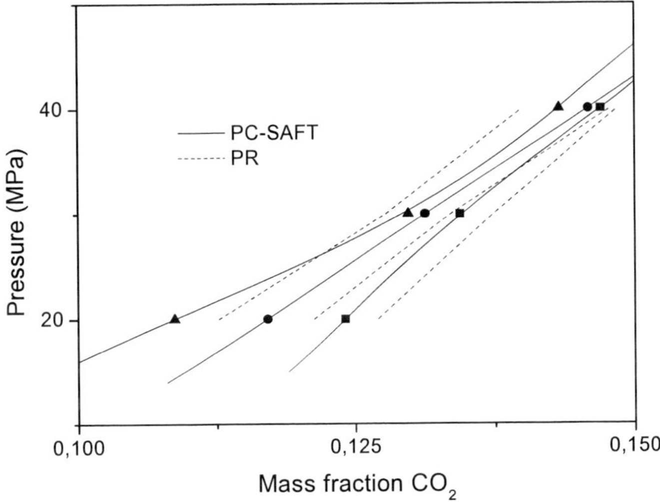

Figure 5.3.9. Solubilities of CO_2 in molten PC. Experimental data (■ = 313.15 K, ● = 323.15 K, ▲= 333.15 K) were taken from Tang et al. (2004).

5.4. Modeling of Binary and Ternary Systems Involving a Biodegradable Polymer, aCopolymer and a Supercritical Fluid

5.4.1. Polymer + Fluid and Copolymer + Fluid Equilibrium

In this section, the high-pressure thermodynamic behavior of eight binary systems and one ternary system including two biodegradable polymers (PLA and PBS), one biodegradable copolymer (PBSA), and six solvents (DME, CO_2, CDFM, DFM, TFM and TFE)[2] was studied. Table 5.4.1 lists some properties of these systems. Pure-component parameters of supercritical fluids, polymers and copolymers, which were calculated as described in sections 5.1.1 and 5.1.2, are shown in Tables 5.4.2 and 5.4.3.

[2] PLA: poly(D,L-lactide), PBS: poly(butyl succinate), PBSA: poly(butyl succinate-co-butylene adipate), DME: dimethyl ether, CDFM: chlorodifluoromethane, DFM: difluoromethane, TFM: trifluoromethane, TFE: 1,1,1,2-tetrafluoroethane.

Table 5.4.1. Some properties of binary and ternary systems studied

System	NP	Temperature (K)	Pressure (MPa)	$Mn_{polymer}$ (g/mol)	Mw/Mn	Ref.
PLA + DME	29	327.15 - 373.35	1.65 - 13.83	2 000, 30 000	-	[a]
PLA + CO_2	19	317.55 - 365.85	20.00 - 142.90	2 750, 60 600, 80 900	1.39, 1.59, 2.00	[b,c]
PLA + CDFM	40	338.15 - 392.75	3.09 - 15.72	2 000, 30 000	-	[d,e]
PLA + DFM	56	303.45 - 375.55	33.00 - 55.55	2 000, 30 000	-	[d]
PLA + TFM	40	303.65 - 373.85	46.05 - 84.75	2 000, 30 000	-	[d]
PLA + TFE	48	302.95 - 373.45	19.18 - 40.15	2 000, 30 000	-	[d]
PBS + CO_2	39	393.15 - 453.15	2.13 - 20.14	29 000	4.82	[f]
PBSA + CO_2	37	393.15 - 453.15	2.08 - 20.13	53 000	3.40	[f]
PLA + DME + CO_2	62	303.05 - 373.35	2.47 - 72.45	30 000	-	[a]

[a] Kuk et al. (2001), [b] Conway et al. (2001), [c] Tom and Debenedetti (1991), [d] Kuk et al. (2002), [e] Lee et al. (2000), [f] Sato et al. (2000a).

Table 5.4.2. Pure-component parameters for supercritical solvents

EoS	Parameters	DME	CO_2	CDFM	DFM	TFM	TFE
	m/MW (mol)$^{-1}$	0.0505	0.0482	0.0243	0.0485	0.0420	0.0194
	σ ($\times 10^{+10}$ m)	3.2302	2.7352	3.3104	2.7852	2.7213	3.1418
PC-SAFT	ε/k (K)	209.4574	166.2143	203.3215	168.4015	141.1332	189.1402
	ΔP^{l} [a]	0.03	0.04	0.41	0.46	0.40	0.58
	Δv^{l} [b]	0.38	0.26	0.40	0.28	0.22	0.31
EoS	**Parameters**	**DME**	**CO_2**	**CDFM**	**DFM**	**TFM**	**TFE**
	T^* (K)	412.12	301.23	389.15	428.22	420.51	350.42
	P^* (MPa)	342.13	585.61	380.22	320.42	380.40	364.42
SL	ρ^* (kg/m^3)	890.21	1532.53	1330.29	1059.45	1108.32	1280.45
	ΔP^{l} [a]	0.28	0.19	0.35	0.45	0.44	0.41
	Δv^{l} [b]	0.69	0.25	0.42	0.32	0.40	0.12

[a] $\Delta P^{l} = \frac{1}{NP}\sum_{i}^{NP}\frac{\left|P_{\exp,i}^{sat} - P_{calc,i}^{satc}\right|}{P_{\exp,i}^{sat}}$, [b] $\Delta v^{l} = \frac{1}{NP}\sum_{i}^{NP}\frac{\left|v_{\exp,i}^{l} - v_{calc,i}^{l}\right|}{v_{\exp,i}^{l}}$

Table 5.4.3. Pure-component parameters for biodegradable polymers and copolymers

EoS	Parameters	PLA	PBS	PBSA
PC-SAFT	m/MW (mol)$^{-1}$	0.3446	0.5138	0.2163
	σ ($\times 10^{+10}$ m)	1.7522	1.4932	1.9911
	ε/k (K)	577.2365	411.3652	399.3601
	ΔP^{L} [a]	1.65	0.97	2.15
	Δv^{L} [b]	2.18	2.51	2.98
SL	T^* (K)	693.12	710.25	688.50
	P^* (MPa)	511.21	531.21	565.82
	ρ^* (kg/m^3)	1232.14	1211.42	1236.48
	ΔP^{L} [a]	2.87	2.45	2.95
	Δv^{L} [b]	3.13	3.85	3.54
PR	a/MW (m^6.MPa/mol^2)	0.2365	0.2124	0.2450
	b/MW ($\times 10^6$ m^3/mol)	1.3215	1.2134	1.2815
	ΔP^{L} [a]	1.10	1.88	2.40
	Δv^{L} [b]	1.15	1.95	3.48

[a] $\Delta P^L = \frac{1}{NP}\sum_{i}^{NP}\frac{\left|P_{\exp,i}^L - P_{calc,i}^L\right|}{P_{\exp,i}^L}$, [b] $\Delta v^L = \frac{1}{NP}\sum_{i}^{NP}\frac{\left|v_{\exp,i}^L - v_{calc,i}^L\right|}{v_{\exp,i}^L}$

Table 5.4.4 shows the temperature-dependent binary interaction parameters for each polymer + supercritical solvent system, which were correlated by Eq. (69).

Table 5.4.4. Temperature-dependent binary interaction parameters for polymer + supercritical solvent systems

System	Binary interaction parameter (κ_{ij})		
	PC-SAFT	SL	PR
PLA + DME	0.0103 + 11.23/T	0.0186 + 4.36/T	0.0522 – 12.84/T
PLA + CO_2	0.0615 – 17.58/T	0.0315 – 6.12/T	-0.0223 + 9.54/T
PLA + CDFM	-0.0128 + 10.45/T	-0.0214 + 15.11/T	-0.0145 + 9.86/T
PLA + DFM	0.0243 – 3.85/T	-0.0189 + 13.15/T	0.0183 – 3.14/T
PLA + TFM	0.0201 + 0.45/T	-0.0182 + 10.85/T	0.0215 – 2.15/T
PLA + TFE	0.0192 – 2.11/T	-0.0095 + 10.45/T	-0.0208 + 14.55/T
PBS + CO_2	0.0085 + 3.45/T	0.0125 + 5.61/T	0.0432 – 21.41/T
PBSA + CO_2	-0.0105 + 12.36/T	-0.0025 – 8.45/T	0.0018 – 5.23/T

Table 5.4.5. Pressure deviations for polymer + supercritical solvent systems

System	ΔP [a]		
	PC-SAFT	SL	PR
PLA + DME	0.75	2.63	4.46
PLA + CO_2	0.86	2.83	4.72
PLA + CDFM	0.82	3.12	3.88
PLA + DFM	0.78	2.95	4.05
PLA + TFM	0.94	2.48	4.95
PLA + TFE	1.03	2.69	5.15
PBS + CO_2	0.72	2.26	3.58
PBSA + CO_2	1.25	3.46	6.40
Global mean deviation[b]	0.90	2.81	4.62

[a] $\Delta P = \frac{1}{NP}\sum_{i}^{NP}\frac{\left|P_{\exp,i} - P_{calc,i}\right|}{P_{\exp,i}}$, [b] $GMD = \frac{\sum_{i}^{NS} NP_i \Delta P_i}{\sum_{i}^{NS} NP_i}$

The impact of this temperature-dependence of κ_{ij} on the cross-segment energy is less than 1.30% for PC-SAFT and 3.50% for SL, in terms of cloud point pressures within the temperature range of experiment for polymer + solvent systems. In Table 5.4.5, global mean deviations are shown in terms of cloud point. It is very clear that PC-SAFT and SL models are more successful than PR in modeling these binary polymer + fluid and copolymer + fluid systems. PC-SAFT EoS has the lowest cloud point pressure deviation (0.90%), SL EoS has 2.81% and PR EoS has the highest deviation in pressure (4.62%). Again, the main difference among the models is that PC-SAFT and SL take in account the molecular structure; this is of vital importance in the case of complex molecules as polymers and copolymers.

For phase equilibrium calculations, it was considered that both the polymer and copolymer are monodisperses ($Mw/Mn \rightarrow 1.00$) (Koningsveld et al., 2001). Binary interaction parameters were determined by fitting selected experimental liquid-fluid equilibrium data by using the modified maximum likelihood method (Niesen and Yesavage, 1989; Stragevitch and d'Avila, 1997) by minimizing the objective function shown in Eq. (68).

5.4.1.1. PLA + DME System

Results obtained in modeling the PLA + DME system (MW_{PLA} = 30 000) in terms of temperature-dependent binary interaction parameters and relative pressure deviations are presented in Tables 5.4.4 and 5.4.5, respectively, for the temperature interval of experimental data supplied by Kuk et al. (2001). PC-SAFT EoS had the best performance, with a deviation smaller than 0.80% in pressure, while the modeling done with SL and PR EoS were not as satisfactory. Figure 5.4.1 shows the pressure isopleths (cloud point curves) against the temperature in the cloud points of PLA (MW = 30 000) in DME at several polymer concentrations (until almost 15.00% in mass) obtained with the three thermodynamic models. In this figure, the DME saturation curve is also shown, which was obtained from DIPPR (2001). The isopleths separate the one-phase region (above) from the two-phase (LL) region (below). This system exhibits characteristics of a phase behavior corresponding to lower critical solution temperature (LCST) phase transition. It is noticed that DME is a good solvent to dissolve polar polymers due its high dipole moment (μ = 1.301 D [DIPPR, 2001]); then it is expected that could dissolve PLA. This is comprobated since the temperature isopleths of phase equilibria (for pressures lower than 14.00 MPa) are lower than DME critical temperature. From Figure 5.4.1 it can also be noticed that cloud point curves have similar inclinations for all polymer concentrations in the mixture. The isopleths cross the DME saturation curve in a lower critical end point (LCEP), in which the last phase transition happens between fluid and two-phase area (LL) (P = 1.20 MPa and T = 330.00 K).

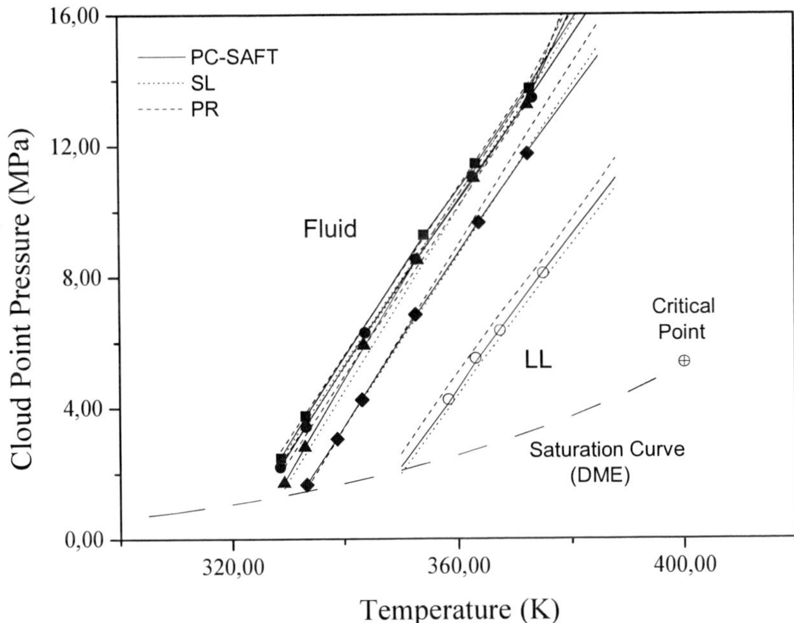

Figure 5.4.1. Cloud point isopleths of PLA + DME systems. Projection of temperature against pressure at different polymer mass percents. Experimental data (○ = 2.97 for MW_{PLA} = 2 000; ◆ = 0.57, ● = 2.91, ■ = 5.00, ▲ = 14.67 for MWPLA = 30 000) were taken from Kuk et al. (2001).

The behavior of pressure against the polymer concentrations for PLA + DME system (MW_{PLA} = 30 000) at constant temperature is shown in Figure 5.4.2. The pressure needed to maintain the polymeric solution in a single phase increases with the temperature. It is also observed that, in each isotherm, the maximum cloud point pressure occurs at constant polymer concentrations in solution (5.00% mass of PLA). At higher concentrations, pressure becomes almost constant, suffering little influence of polymer concentration.

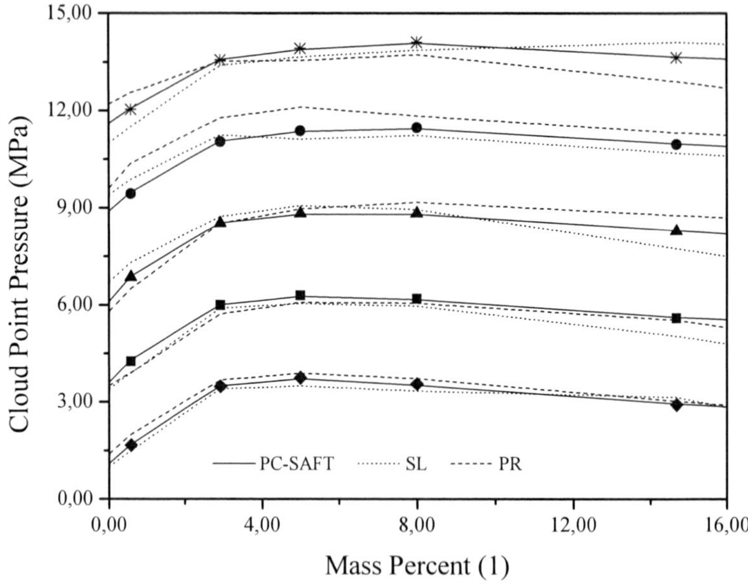

Figure 5.4.2. Pressure vs polymer mass percent isotherms for PLA (MW = 30 000) + DME systems. Experimental data (◆ = 333.15 K, ■ = 343.15 K, ▲ = 353.15 K, ● = 363.15 K, ж = 373.15 K) were taken from Kuk et al. (2001).

5.4.1.2. PLA + CO_2 System

Solubilities of PLA in CO_2 were modeled based on literature data, supplied by Conway et al. (2001) and Tom and Debenedetti (1991). There were used two forms of PLA: Mw/Mn = 1.39 and 1.59 (Conway et al., 2001) and 2.00 (Tom and Debenedetti, 1991), and it can be assumed that, under these conditions, PLA has the bevahior of a monodisperse polymer (Koningsveld et al., 2001). Figure 5.4.3 displays the impact of polydispersivity index (Mw/Mn relationships) on the phase behavior of PLA + CO_2 system. Cloud points exhibit a slight positive inclination. In a similar way to Figure 5.4.1, this binary system also presents a phase behavior of LCST type, but with some differences that it is important to write down. For instance, higher pressures are necessaries (higher than 100.00 MPa) for dissolving PLA in CO_2.; in this manner, a final transition region between the fluid phase and LL phase (occurrence presented in Figure 5.4.1) on the intersection of isopleths with the CO_2 saturation curve is not possible for this binary system, due to mainly to the dipole moment of CO_2 (μ = 0,00 D [DIPPR, 2000]) in contrast to the dipole moment of DME (μ = 1,30 D [DIPPR, 2000]). At lower relative pressures and temperatures, CO_2 it is unable to dissolve to PLA. The dependency of PLA mass percents with cloud point pressures in modeling PLA + CO_2 systems at different temperatures was also studied (Mw/Mn = 2.00) as shown in Figure 5.4.4. It could be noticed that solubilities of PLA in CO_2 increase when cloud point pressures and temperatures increase. At 338.15 K, the relationships between solubilities of PLA in CO_2 with pressure are linear, while at lower temperatures, the relationships are linear until pressures up to 25.00 MPa, with almost horizontal slopes, which are more pronounced at pressures higher than 25.00 MPa to favor solubilities of PLA in CO_2.

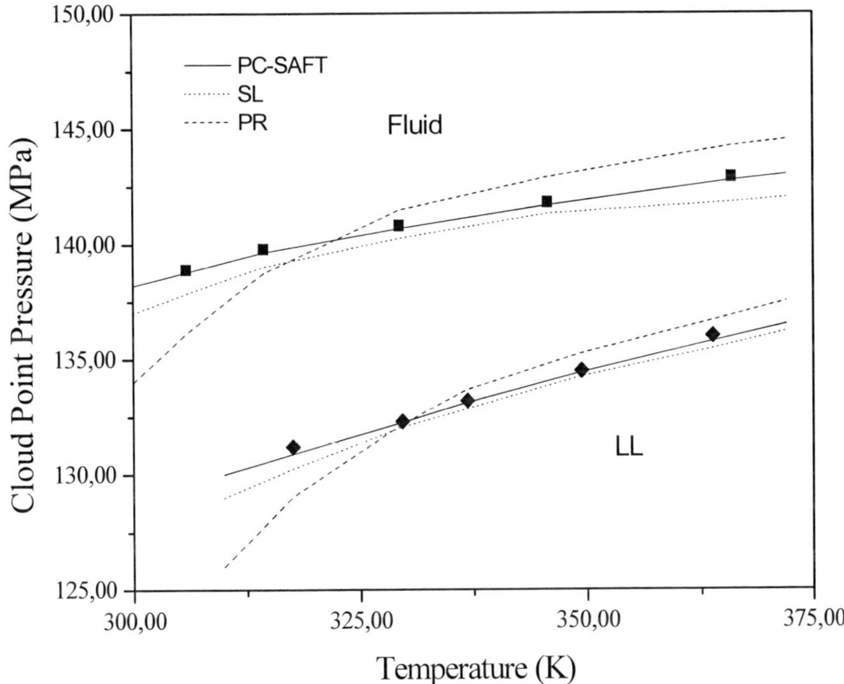

Figure 5.4.3. Influence of polydispersivity index (Mw/Mn) in pressure - temperature projections in modeling binary systems of PLA + CO_2. Experimental data (◆ = 1.39 and ■ = 1.59) were taken from Conway et al. (2001).

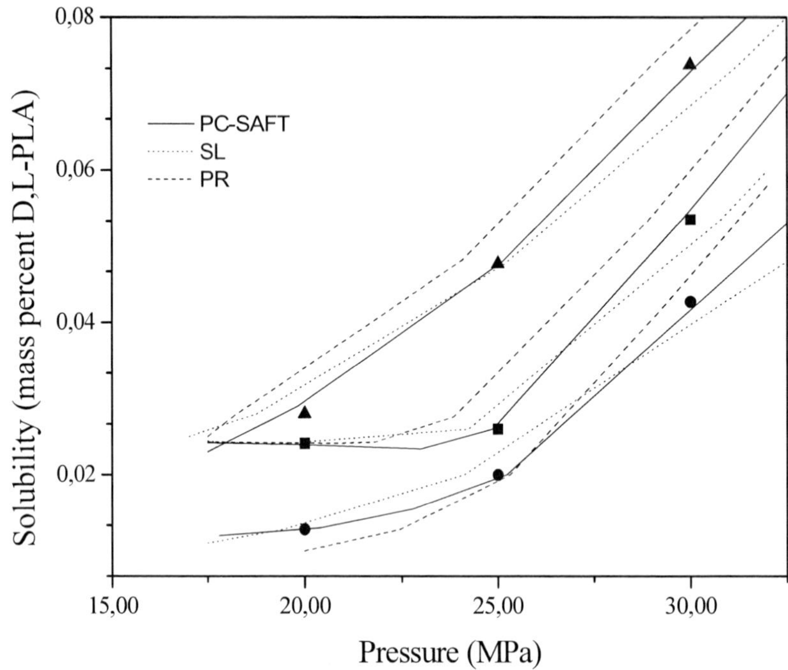

Figure 5.4.4. Solubilities of PLA in supercritical CO_2 at several temperatures. Experimental data (●= 318.15 K, ■ = 325.15 K, ▲= 338.15 K) were taken from Tom and Debenedetti (1991).

5.4.1.3. PLA + CDFM System

The phase behavior of PLA + CDFM systems for two different molecular weights with similar polymer mass percents (2.74 and 5.78 for MW_{PLA} = 2 000 and 2.87 and 4.77 for MW_{PLA} = 30 000) is compared and shown in Figure 5.4.5. For all the systems examined, the LCST curve intersects with the CDFM saturation curve and no phase behavior of U-LCST type is detected. As for the PLA + DME system, increasing the polymer molecular weight decreases the one-phase region. PC-SAFT results are in a very good agreement with the experimental data, while the other models (SL and PR) can also correlate the phase behavior of this binary system. As for DME, only one temperature-dependent binary parameter is used.

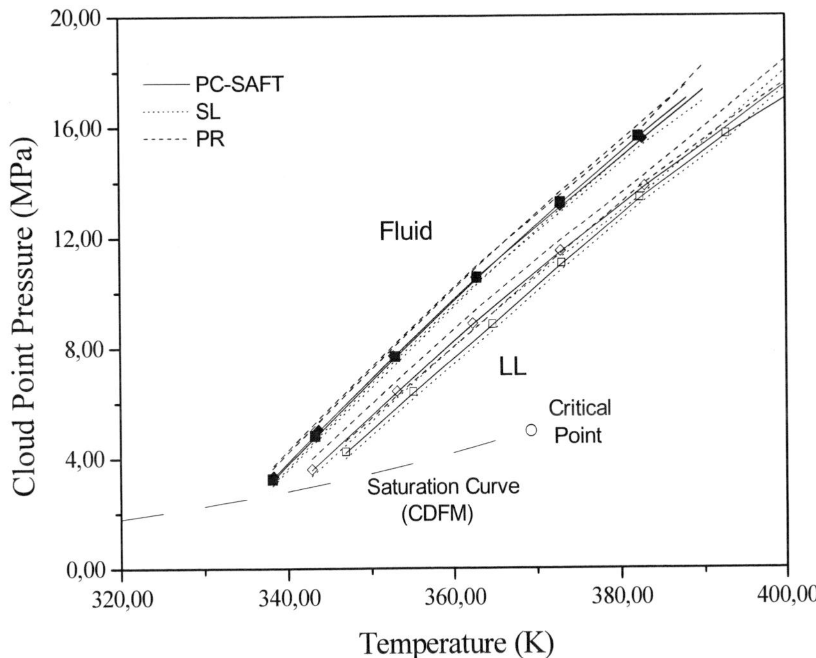

Figure 5.4.5. Cloud point isopleths of PLA (MW = 30 000) + CDFM systems. Projection of temperature against pressure at different polymer mass percents (□ = 2.74, ○ = 5.68 for MWPLA = 2 000; ■ = 2.87, ● = 4.77). Experimental data were taken from Conway et al. (2001), Kuk et al. (2002) and Lee et al. (2000).

Figure 5.4.6 shows the phase behavior of the PLA + CDFM system for different polymer mass percents (MW_{PLA} = 30 000). It is possible to see the intersection of the LCST curve with the CDFM saturation curve for all curves correlated with the PC-SAFT, SL and PR models. It is important to stand out that, when the polymer mass percent increases, the one-phase region decreases. Several isotherms of cloud point pressures against the polymer mass percents for PLA + CDFM systems (MW_{PLA} = 30 000) were also modeled and appear in Figure 5.4.7. The cloud point pressures did not vary much with the polymer concentration; the isotherms are almost horizontal. In each isotherm, the maximum cloud point pressure calculated for each thermodynamic model, which corresponds to an upper critical solution pressure (UCSP), was in agreement with the experimental data at the polymer concentration around 3.00 or 4.00 mass percent.

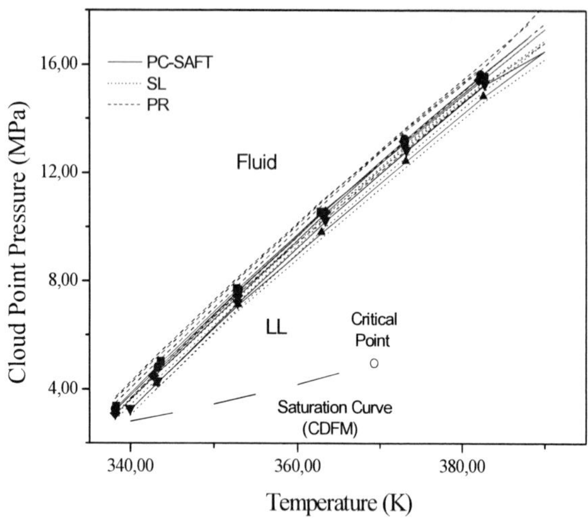

Figure 5.4.6. Cloud point isopleths of PLA (MW = 30 000) + CDFM systems. Projection of temperature against pressure at different polymer mass percents. Experimental data (▲ = 0.50, ■ = 2.87, ● = 4.77, ◆ = 7.84 and ▼ = 14.68 for) were taken from Kuk et al. (2002) and Lee et al. (2000).

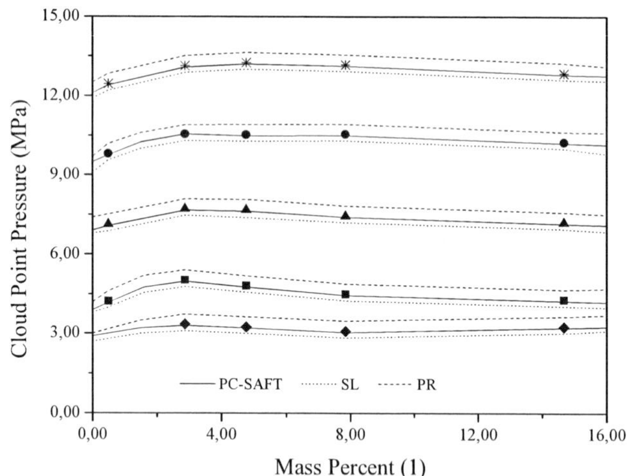

Figure 5.4.7. Cloud point vs polymer mass percent isotherms for PLA (MW = 30 000) + CDFM systems. Experimental data (◆ = 338.63 K, ■ = 343.19 K, ▲ = 352.85 K, ● = 363.03 K, ж = 373.00 K) were taken from Kuk et al. (2002) and Lee et al. (2000).

5.4.1.4. PLA + DFM System

As the molecular weight of the polymer increases (from systems of MW_{PLA} = 2 000 containing 3.22 mass percent of polymer to systems of MW_{PLA} = 30 000 containing 2.78 mass percent of polymer), the region of complete miscibility (above the LCST boundary) shrinks, merging for PLA + DFM systems in a U-LCST curve, as shown in Figure 5.4.8. While PC-SAFT model correlates the U-LCST curve with good precision, the other models (SL and PR) can also describe the phase behavior of this binary system. In Figure 5.4.8, the phase behavior of PLA + DFM systems, for different polymer mass percents (MW_{PLA} = 30 000), is also shown, correlated with the PC-SAFT, SL and PR models. It is easy to observe the presence of the U-LCST minimum for all polymer mass percents above the DFM saturation curve. The correlation of experimental data with the PC-SAFT model is the best one.

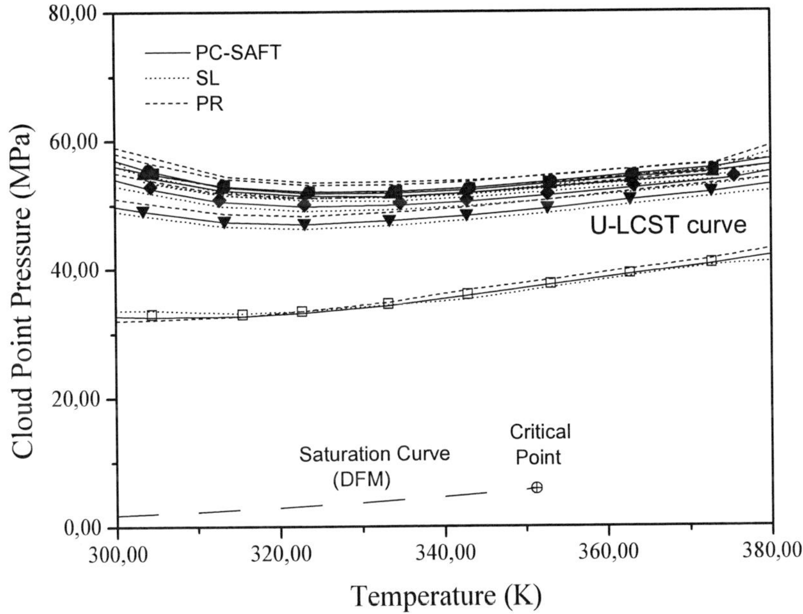

Figure 5.4.8. Cloud point isopleths of PLA + DFM systems. Projection of temperature against pressure at different polymer mass percents. Experimental data (□ = 3.22 for MW_{PLA} = 2 000; ▲ = 1.64, ■ = 2.78, ● = 4.80, ◆ = 9.08 and ▼ = 15.00 for MW_{PLA} = 30 000) were taken from Kuk et al. (2002).

The dependency of cloud point pressures against the polymer mass percents for the PLA + DFM system (MW_{PLA} = 30 000) at constant temperature was also analyzed and shown in Figure 5.4.9. As the polymer concentration increases, the cloud point pressure increases, achieving a maximum at polymer concentrations of 3.00 to 4.00 mass percent, and then gradually decreasing. PC-SAFT, SL and PR EoS can model the phase behavior of this binary system for all the isotherms.

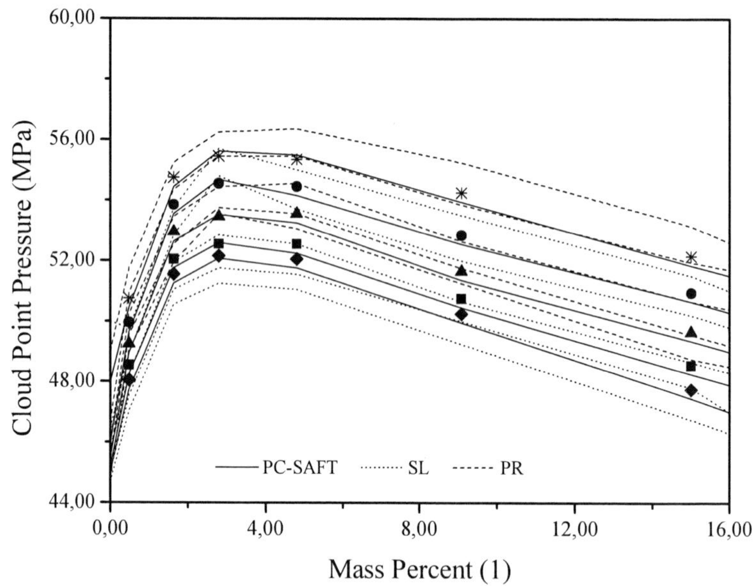

Figure 5.4.9. Cloud point vs polymer mass percent isotherms for PLA (MW = 30 000) + DFM systems. Experimental data (◆ = 333.98 K, ■ = 343.10 K, ▲ = 353.00 K, ● = 363.13 K, Ж = 373.38 K) were taken from Kuk et al. (2002).

5.4.1.5. PLA + TFM System

The phase behavior of PLA in TFM in terms of cloud point pressures is modeled by the PC-SAFT, SL and PR EoS. In Figure 5.4.10, for systems with two molecular weights of PLA (MW_{PLA} = 2 000 and 30 000) with similar polymer mass percents (3.06 and 3.10), cloud point pressures increase with increasing temperatures, indicating that these systems also exhibit the characteristics of an LCST phase behavior. As for other binary systems studied, if the polymer molecular weight increases, the cloud point pressure also increases at constant temperature. Figure 5.4.10 also shows the pressure vs. temperature experimental isopleths for different polymer mass concentrations for MW_{PLA} = 30 000 and the isopleths obtained from the PC-SAFT, SL and PR models. When the polymer mass percent increases (at constant molecular weight), the cloud point pressure also increases at constant temperature. It is important to notice that these isopleths also have LCST phase behavior and the intersection with the TFM saturation curve will occur at very low temperatures.

Cloud point pressure vs. polymer mass percent isotherms of PLA in TFM (MW_{PLA} = 30 000) at several temperatures were also studied; the isotherms are shown in Figure 5.4.11. From this study, it can be concluded that the maximum cloud point pressures are observed at the polymer concentration around 8.0 mass percent.

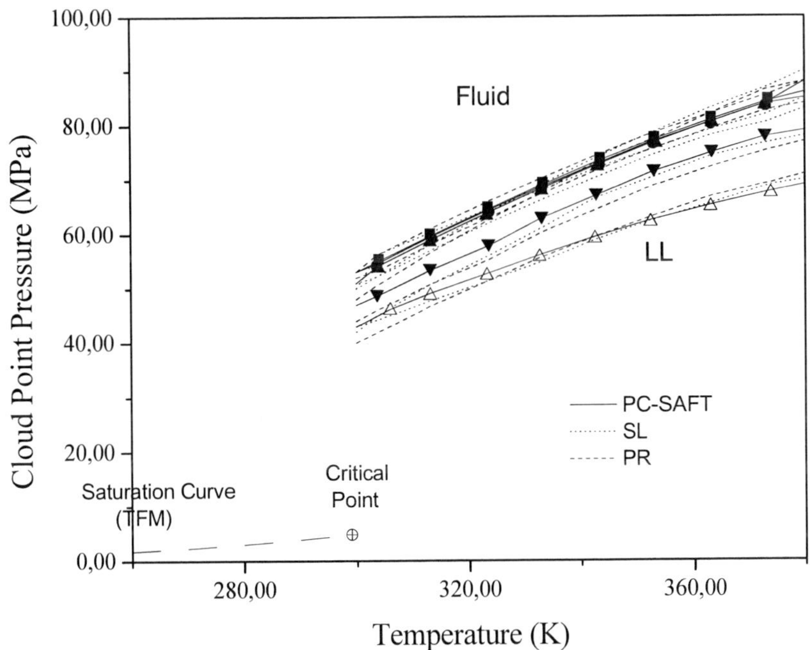

Figure 5.4.10. Cloud point isopleths of PLA + TFM systems. Projection of temperature against pressure at different polymer mass percents. Experimental data (△ = 3.06 for $MW_{PL}A$ = 2 000; ▼ = 0.55, ▲ = 3.10, ■ = 7.93 and ● = 13.17 for MW_{PLA} = 30 000) were taken from Conway et al. (2001) and Kuk et al. (2002).

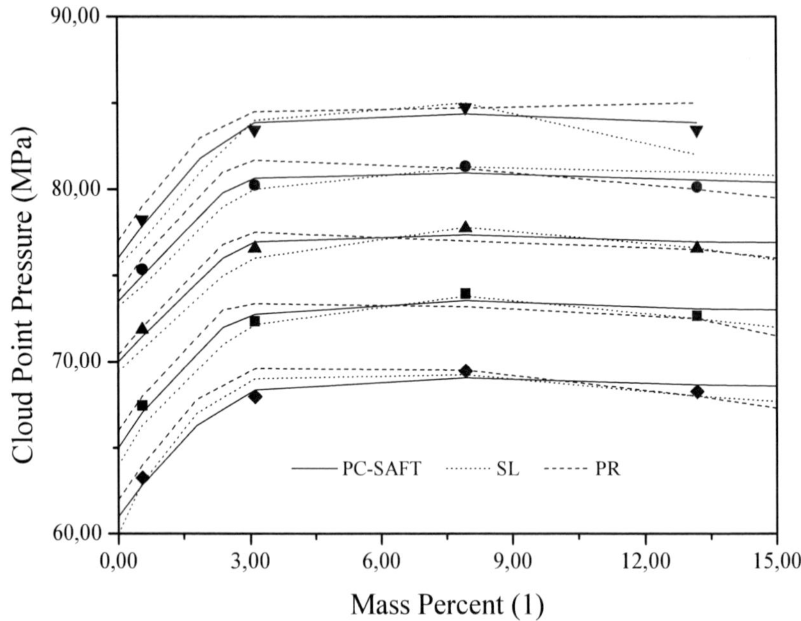

Figure 5.4.11. Pressure vs polymer mass percent isotherms for PLA (MW = 30 000) + TFM systems. Experimental data (◆ = 333.20 K, ■ = 343.25 K, ▲ = 353.25 K, ● = 363.23 K, Ж = 373.00 K) were taken from Conway et al. (2001) and Kuk et al. (2002).

5.4.1.6. PLA + TFE System

Cloud point experimental data (Kuk et al., 2002) of PLA in TFE are modeled with PC-SAFT, SL and PR EoS. In Figure 5.4.12, the cloud point pressure increases with increasing temperature, indicating that these systems also exhibit the characteristics of an LCST phase behavior. In this figure, the results obtained by the three thermodynamic models are compared with the experimental data for two polymer molecular weights (MW_{PLA} = 2 000 and 30 000) with similar polymer mass percents in each mixture (2.92 and 3.04). At higher polymer molecular weights, the cloud point pressures increases at constant temperature. Modeling is also shown for several mixtures with different polymer mass percents in a cloud point pressure vs. temperature diagram for MW_{PLA} = 30 000. When the polymer mass percent increases, the cloud point pressure also increases at constant temperature. It is noticed that the fluid + LL isopleths (LCST curves) will intersect the TFE saturation curve at lower temperatures.

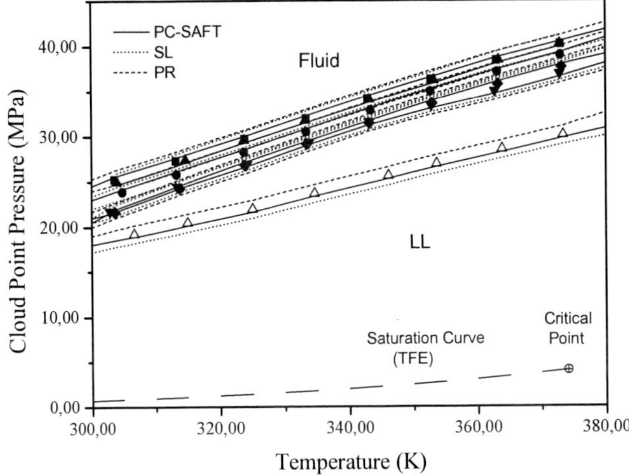

Figure 5.4.12. Cloud point isopleths of PLA + TFE systems. Projection of temperature against pressure at different polymer mass percents. Experimental data (△ = 2.92 for MW_{PLA} = 2 000; ▼ = 0.54, ▲ = 3.04, ■ = 5.26 and ● = 8.55 and ◆ = 15.05 for MW_{PLA} = 30 000) were taken from Kuk et al. (2002).

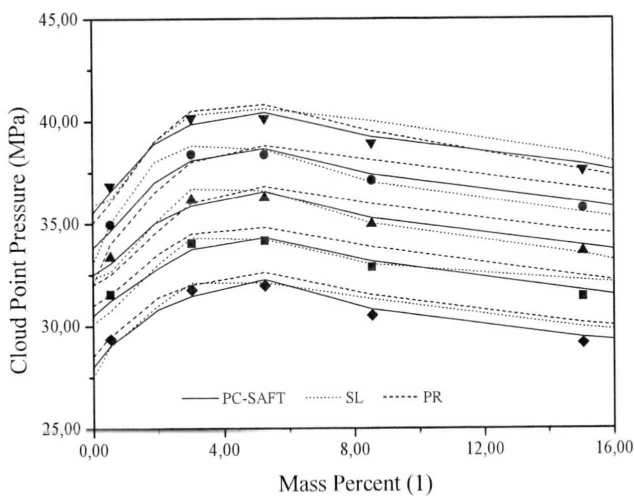

Figure 5.4.13. Cloud point vs polymer mass percent isotherms for PLA (MW = 30 000) + TFE systems. Experimental data (◆ = 333.45 K, ■ = 343.15 K, ▲ = 352.93 K, ● = 363.11 K, ж = 373.05 K) were taken from Kuk et al. (2002).

Isotherms of cloud point pressure against polymer mass percent of PLA (MW = 30 000) in TFE were modeled by analyzing the pressure vs. temperature data as a function of polymer mass percents at various temperatures, as shown in Figure 5.4.13. The maximum cloud point pressures are observed at the polymer concentration between 4.00 and 6.00 mass percent.

5.4.1.7. PBS + CO_2 System

Solubilities for CO_2 in PBS (Mw/Mn = 4.83) (Sato et al., 2000a) are studied in terms of cloud point pressure against fluid mass percent at different temperatures, and in terms of cloud point pressure against temperature at different fluid mass percents. The results are shown in Figures 5.4.14 and 5.4.15, respectively; in both cases, the data were modeled with the PC-SAFT, SL and PR EoS. Fluid mass percents vary in an almost linear form with the pressure at a given temperature, with the slope changing when temperature increases, as shown in Figure 5.4.15.

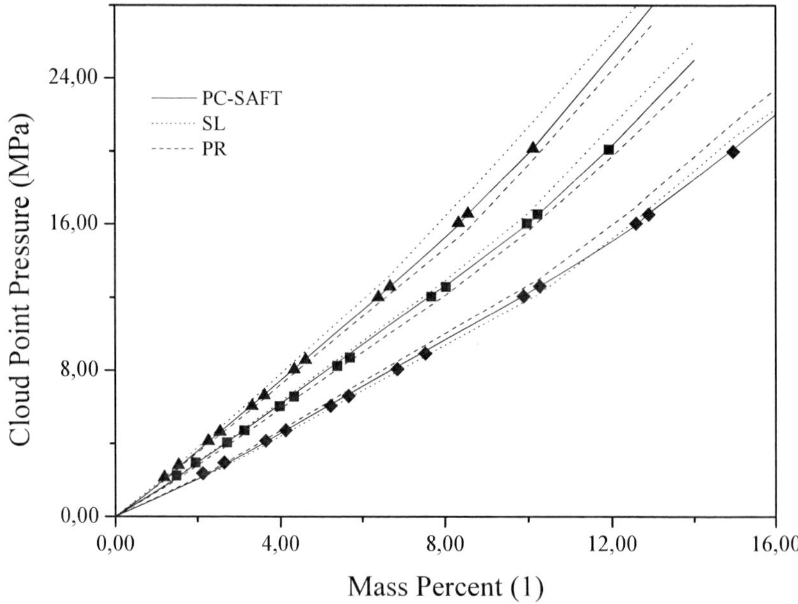

Figure 5.4.14. Pressure-polymer concentration isotherms for MW_{PBS} = 29 000 (Mw/Mn = 4.83) in CO_2 at various temperatures. Experimental data (◆ = 393.15 K, ■ = 423.15 K, ▲ = 453.15 K) were taken from Sato et al. (2000a).

In Figure 5.4.14, the same behavior is observed, but the pressure vs. temperature slope at a given fluid mass percents changes more slowly. In the same figure, it is important to see that, if the cloud point pressure remains constant, an increasing of temperature causes a decreasing of fluid mass percent; in other words, the solubility of fluid in polymer decreases.

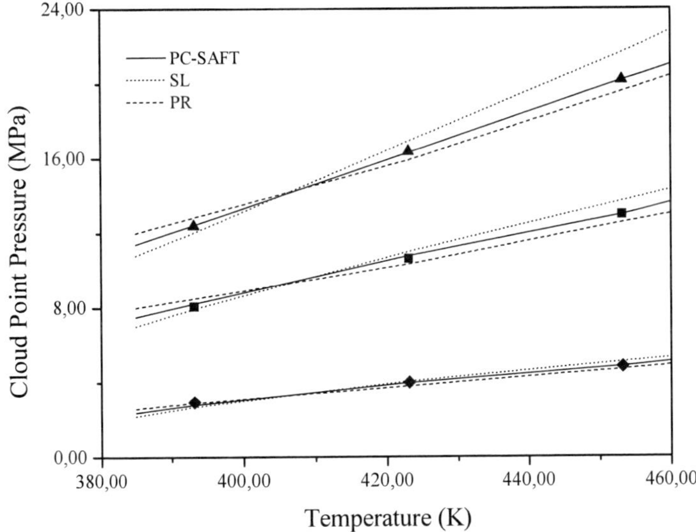

Figure 5.4.15. Cloud point isopleths of PBS (MW = 29 000) + CO_2 system. Projection of temperature against pressure at different fluid mass percents. Experimental data (◆ = 2.64%, ■ = 6.85% and ▲ = 10.12%) were taken from Sato et al. (2000a).

5.4.1.8. PBSA + CO_2 System

Fluid mass percent isopleths are shown in Figure 5.4.16 for the PBSA + CO_2 system. When the fluid mass percent is higher, the positive slope increases. Isotherms originated by modeling the experimental data (Sato et al., 2000a) of solubilities of CO_2 in PBSA (Mw/Mn = 3.39) are also modeled and appear in Figure 5.4.17. At constant temperature, the fluid mass percent varies in almost linear form with the system cloud point pressure. At all temperatures, the PC-SAFT EoS can model the phase behavior of this binary system with more precision than the SL and PR EoS; even so, the latter models show the trend of the experimental data, but with higher deviations in terms of cloud point pressure.

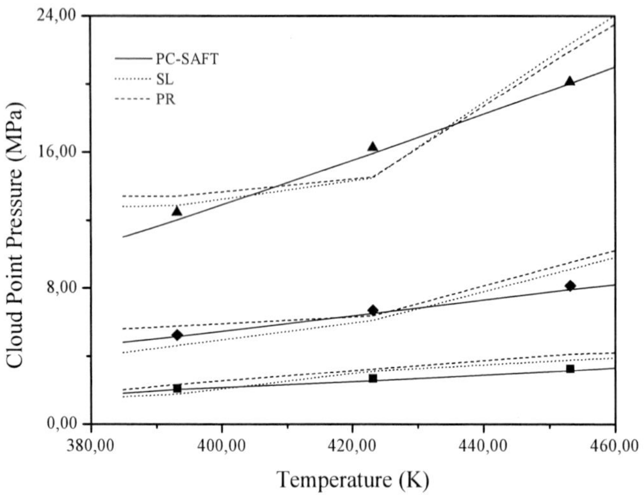

Figure 5.4.16. Cloud point isopleths of PBSA (MW = 53 000) + CO_2 system. Projection of temperature against pressure at different fluid mass percents: ■ = 1.81%, ◆ = 4.55% and ▲ = 10.20%. Experimental data were taken from Sato et al. (2000a).

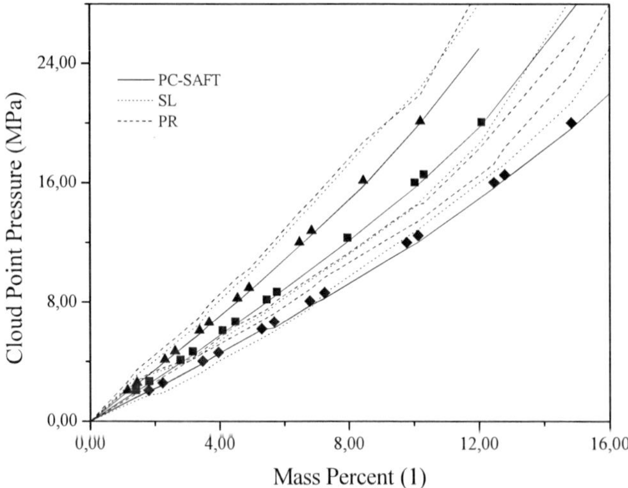

Figure 5.4.17. Pressure-polymer concentration isotherms for MW_{PBSA} = 53 000 (Mw/Mn = 3.39) in CO_2 at various temperatures. Experimental data (◆ = 393.15 K, ■ = 423.15 K, ▲ = 453.15 K) were taken from Sato et al. (2000a).

5.4.2. Polymer + Fluid 1 + Fluid 2 Equilibrium

5.4.2.1. PLA + CO_2 + DME System

Experimental data of cloud points (Kuk et al., 2001) are compared with those obtained by the PC-SAFT, SL and PR models in the prediction of phase behavior in ternary systems containing PLA in DME + CO_2 mixtures. This ternary system was studied from its three binary systems (PLA + CO_2, PLA + DME and DME + CO_2).

A comparison between experimental data of DME + CO_2 system (Jónasson et al., 1995; Laursen et al., 2002) and values calculated by thermodynamic models using classic mixing rules with binary interaction parameters ($\kappa_{DME+CO2}$) by fitting experimental values was made by Arce (2001). The values calculated by the PC-SAFT EoS are in good agreement with experimental data, as shown in Figure 5.4.18. The curves obtained with the other models (SL and PR) present a slight deviation when compared in a qualitative way with the vapor phase experimental data. In a quantitative view, pressure deviations were 0.69, 1.86 and 2.50% for the PC-SAFT, SL and PR EoS, respectively, and 0.80, 3.08 and 2.66% in vapor phase mole fraction deviations for the same models, respectively. PC-SAFT was also able to predict the critical points.

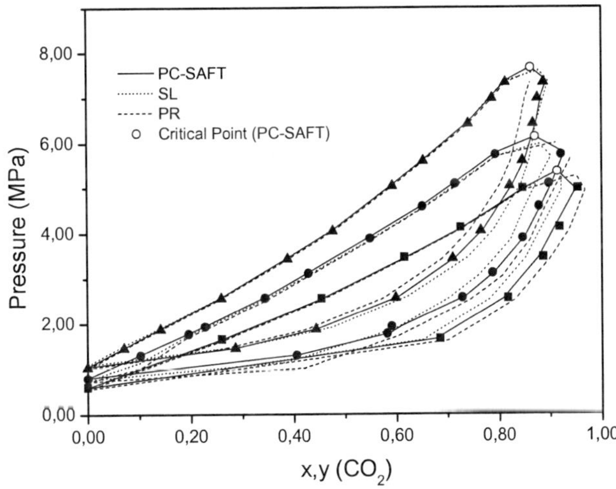

Figure 5.4.18. Modeling of the fluid phase behavior of DME + CO_2 systems. Experimental data (■ = 298.15 K, ● = 308.65 K, ▲ = 320.15 K) were taken from Jónasson et al. (1995) and Laursen et al. (2002).

The binary interaction parameters used in prediction of the phase behavior of this ternary system were the same obtained in the modeling of the binary systems ($\kappa_{PLA + CO2}$, $\kappa_{PLA + DME}$) and. $\kappa_{DME + CO2}$). Figures 5.4.19, 5.4.20 and 5.4.21 show the results obtained with each thermodynamic model in terms of cloud point isopleths of PLA (MW = 30 000) in DME + CO_2 mixtures for several compositions of CO_2 until 73.00% (on a polymer-free basis). Polymer concentration in solution is fixed at 4.76% (mass) in order to eliminate the effect of polymer concentration on the cloud point. Cloud pressure deviations obtained by the PC-SAFT, SL and PR EoS are 0.80, 2.87 and 3.99%, respectively. These deviations were calculated in according with Eq. (33). Figure 5.4.19 displays the effect of CO_2 compositions in solvent mixtures (CO_2 + DME) in terms of cloud point pressures for different temperatures. Cloud point pressures increase abruptly when CO_2 concentration in solvent mixtures increases.

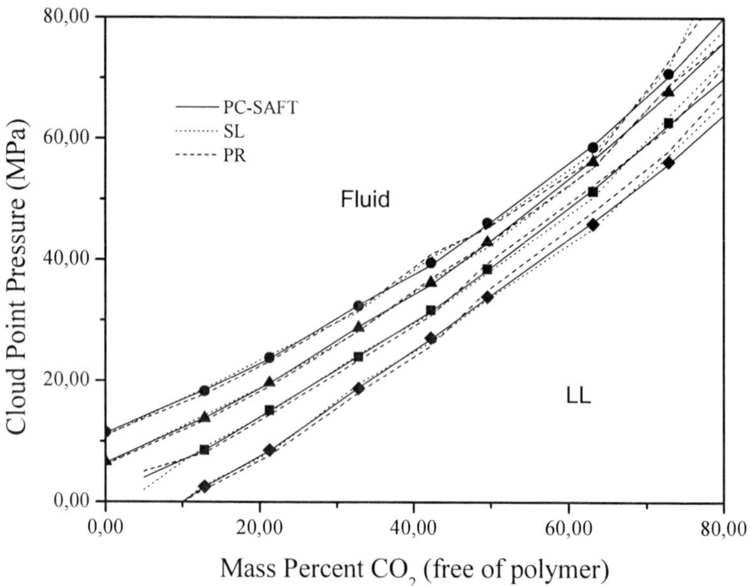

Figure 5.4.19 Composition effect of CO2 in solvent mixtures (DME + CO_2) on cloud point pressures of PLA (MW_{PLA} = 30 000) at several temperatures (◆ = 303.15 K, ■ = 323.15 K, ▲ = 343.15 K and ● = 363.15 K). Experimental data were taken from Kuk et al. (2001).

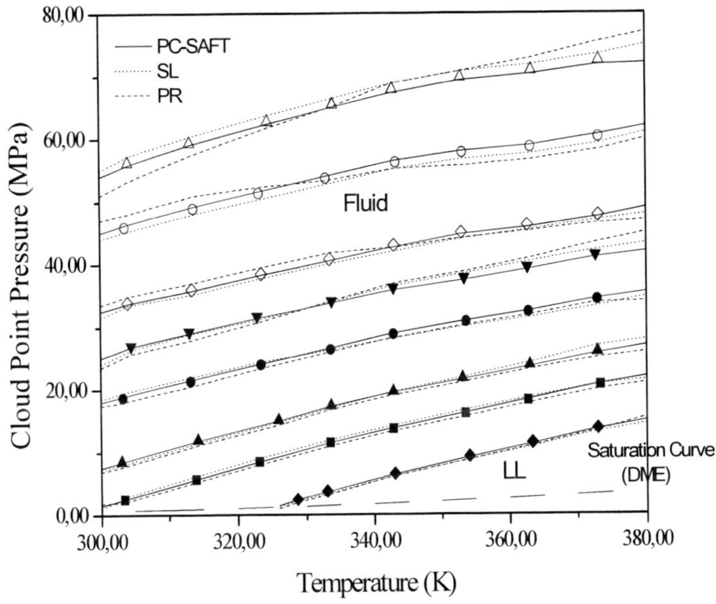

Figure 5.4.20. Cloud point isopleths of PLA (MW_{PLA} = 30 000) in solvent mixtures (CO_2 + DME) at different CO_2 mass percents (◆ = 0.00, ■ = 12.83, ▲ = 21.27, ● = 32.80, ▼ = 42.22, ◇ = 49.55, ○ = 63.10 and △ = 72.79). Experimental data were taken from Kuk et al. (2001).

The cloud point isopleths exhibit the characteristics of a LCST curve for all CO_2 compositions. When the CO_2 concentration in solvent (DME) increases, the cloud point isopleths shift to higher pressures, which causes a shrink of the one-phase region, as shown in Figure 5.4.20. Addition of CO_2 to DME can cause a decrease of dissolution force of solvent mixtures. This can be attributed to the decrease of solvent polarity by increasing of CO_2 composition in solvent.

Figure 5.4.21 displays the effect of CO_2 addition in PLA + DME mixtures by fixing the polymer molecular weight. A ternary mixture formed by PLA, CO_2 and DME presents higher cloud point pressures than the ones obtained by a binary mixture of PLA + DME when the temperature remains constant, obtaining a larger two-phase area (LL) between the cloud point isopleths and the DME saturation curve.

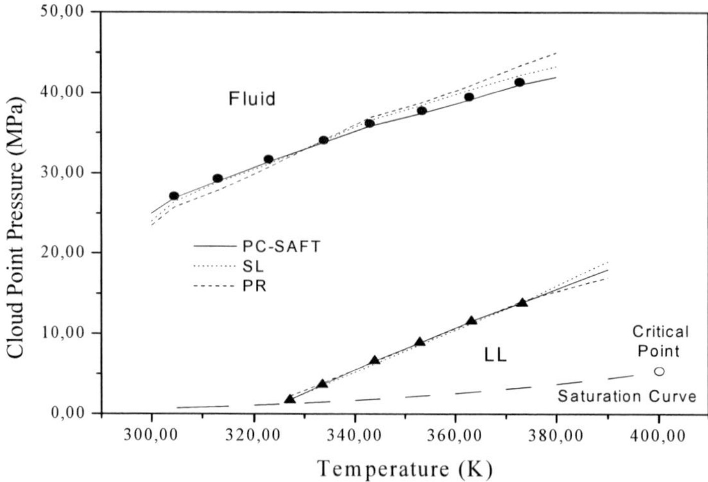

Figure 5.4.2.1. Influence of CO_2 addition in binary mixtures (PLA + DME) on cloud point pressures (▲ = binary mixture and ● = ternary mixture). Experimental data were taken from Kuk et al. (2001).

5.5. Biodegradable Copolymer + Supercritical Fluid Systems

In this section, the fluid phase behavior at high pressure of four binary systems including one biodegradable copolymer (PLAG) with different molecular weights, and four supercritical solvents (DME, CO_2, CDFM and TFM)[1] was studied. EoS pure-component parameters of supercritical solvents were shown in section 5.4.1, while the copolymer pure-component parameters were obtained using the Elvassore et al. (2002) method, as explained in section 5.1.2. Some copolymer physical characteristics and PHSC pure-component parameters appear in Table 5.5.1, while the PC-SAFT, SL and PR pure-component parameters for the PLAG are shown in in Table 5.5.2. These parameters were obtained in a pressure and temperature interval commonly used in engineering calculations.

Table 5.5.3 shows some physical characteristics (pressure and temperature range and experimental data points) of several PLAG + supercritical solvent,

1 PLAG: poly(d,l-lactide-co-glycolide), DME: dimethyl ether, CDFM: chlorodifluoromethane, TFM: trifluoromethane.

while Table 5.5.4 lists the temperature-dependent binary parameters used in modeling these binary systems for the PC-SAFT, SL and PR EoS. The interaction binary parameters were calculated in function of two empirical variables (C_1 and C_2) and were correlated in the form of Eq. (69).

For phase equilibrium calculations of these copolymer + supercritical solvent systems it was considered that only the supercritical fluid is present in vapor phase and that the copolymer tends to be monodisperse (Mw/Mn → 1.00 [Koningsveld et al., 2001]). Binary interaction parameters were obtained by fitting selected experimental liquid-fluid equilibrium data by using the modified maximum likelihood method (Niesen and Yesavage, 1989; Stragevitch and d'Avila, 1997) by minimizing the objective function shown in Eq. (68).

In Table 5.5.4 are also shown the deviations in cloud point pressures obtained between the experimental data and the calculated data for each model thermodynamic. Cloud point pressures are predicted quantitatively over the entire temperature range for each PLAG copolymer + SCF system. It is very clear that the PC-SAFT and SL models are more successful in modeling these binary systems. Again, the greater advantage of these models to represent copolymer + fluid systems is that they take in account the molecular structure. This is the vital importance in the case of complex molecules as copolymers.

For the ten PLAG copolymer + SCF systems described in Table 5.5.3, the calculated PC-SAFT global mean deviation was 1.27%, while the values for the SL and PR models were 3.45 and 4.56%, respectively. From these results, it can be observed a better performance of PC-SAFT EoS over the other thermodynamic models; again, this is due to the more rigorous account of molecular structure and interactions.

5.5.1. PLAG + DME System

The results in terms of cloud point pressure deviations and temperature-dependent binary interaction parameters of PLAG copolymers in DME (μ = 1.301 D [DIPPR, 2000]) at high pressures are given in Table 5.5.4 and Figure 5.5.1. The content of DME in PLAG copolymer + DME systems was between 96.86 and 97.25 mass percent and the D,L-lactide : glycolide mole ratios in the copolymers varied from 87.88 : 12.12 to 49.60 : 50.40. When the D,L-lactide : glycolide mole ratio decreased, the cloud point pressures increased at a given temperature.

Table 5.5.1. PLAG properties and PHSC pure-component parameters

D,L-lactide/glycolide monomer molar ratio	PLAG copolymer	Polidispersity (Mw/Mn)	M_η (kg/mol)	PHSC pure-component parameters		
				A^* 10^8 (m^2/mol)	V^* 10^3 (m^3/mol)	E^* (MPa m^3/mol)
87.88 / 12.12	1	[a]	5.00	10.1210	32.4515	6.5044
85.00 / 15.00	2	1.73	[b]	8.6065	29.3014	6.2520
85.00 / 15.00	3	1.54	[b]	8.4018	28.1008	6.1810
76.31 / 23.69	4	[a]	5.00	4.5025	16.6025	3.4718
75.00 / 25.00	5	1.67	[b]	4.2810	15.1410	3.5014
65.27 / 34.73	6	[a]	10.00	2.5918	8.8515	2.1018

[a] Polidispersitiy index is not available, [b] viscosity average molecular weight is not available.

Table 5.5.2. PLAG pure-component parameters for PC-SAFT, SL and PR EoS

EoS	Parameters	PLAG copolymer					
		1	2	3	4	5	6
PC-SAFT	m/MW (mol)$^{-1}$	0.3228	0.3345	0.3440	0.3885	0.3940	0.4415
	σ ($\times 10^{+10}$ m)	1.8018	1.7926	1.8043	1.7515	1.7329	1.7015
	ε/k (K)	590.25	592.14	593.10	570.14	568.15	549.52
	$\Delta P^{L\,a}$	1.29	1.53	1.43	1.49	1.51	1.28
	$\Delta v^{L\,b}$	2.18	2.30	2.25	2.36	2.45	2.29
SL	T* (K)	698.15	696.92	695.04	685.17	686.90	670.42
	P* (MPa)	530.33	532.20	535.43	525.22	523.18	512.10
	ρ^* (Kg/m^3)	1432.01	1425.15	1435.47	1350.51	1362.83	1320.12
	$\Delta P^{L\,a}$	3.15	3.21	3.05	4.12	3.83	3.93
	$\Delta v^{L\,b}$	4.18	4.01	4.19	4.39	4.53	4.14
PR	a/MW (m^6.MPa/mol^2)	0.3315	0.3298	0.3363	0.2919	0.2965	0.2514
	b/MW *10^6 (m^3/mol)	1.5120	1.5029	1.5156	1.4503	1.4585	1.3901
	$\Delta P^{L\,a}$	3.02	3.86	3.51	3.89	4.23	4.20
	$\Delta v^{L\,b}$	4.28	4.95	5.26	5.28	5.09	5.14

[a] $\Delta P^L = \frac{1}{NP}\sum_i^{NP}\frac{\left|P^L_{exp,i} - P^L_{calc,i}\right|}{P^L_{exp,i}}$, [b] $\Delta v^L = \frac{1}{NP}\sum_i^{NP}\frac{\left|v^L_{exp,i} - v^L_{calc,i}\right|}{v^L_{exp,i}}$

Table 5.5.3. Main physical characteristics of binary systems studied in this work

Binary system		NP	Temperature (K)	Pressure (MPa)	Ref
SCF	PLAG copolymer				
DME +	1	8	303.65 - 373.25	8.55 - 23.29	[a]
	4	8	303.35 - 373.05	21.65 - 32.65	
	6	8	305.15 - 373.45	44.25 - 49.05	
CO_2 +	2	5	309.85 - 358.65	184.30 - 191.80	[b]
	3	5	312.15 - 364.85	177.00 - 182.20	
	5	5	312.05 - 352.95	219.80 - 239.40	
CDFM +	1	6	331.85 - 383.15	3.73 - 19.67	[c]
	4	7	325.05 - 382.55	3.45 - 22.60	
	6	8	314.65 - 382.55	3.80 - 29.65	
TFM +	5	5	301.75 - 354.25	134.10 - 147.40	[b]

[a] Kuk et al. (2001), [b] Conway et al. (2001), [c] Lee et al. (2000).

Table 5.5.4. Temperature-dependent binary interaction parameters and pressure deviations of binary systems studied in this work

Binary system		Equation of State					
		PC-SAFT		SL		PR	
SCF	PLAG copolymer	κ_{ij}	ΔP [a]	κ_{ij}	ΔP [a]	κ_{ij}	ΔP [a]
DME +	1	-0.0182 + 12.4528/T	1.97	0.5263 – 185.7458/T	3.59	-0.0125 – 10.2155/T	8.40
	4	-0.0261 + 13.5361/T	1.66	0.3896 – 141.7463/T	2.60	-0.0311 – 8.2320/T	4.50
	6	-0.0383 + 11.6952/T	1.68	0.3012 – 118.2332/T	2.20	0.0113 – 10.3212/T	2.09
CO_2 +	2	0.0195 – 5.3623/T	0.42	-0.2513 + 92.5896/T	2.59	-0.5863 + 208.2363/T	1.01
	3	0.0121 – 1.4120/T	0.51	-0.2228 + 83.1521/T	1.11	-0.7841 + 272.4150/T	1.17
	5	-0.0035 – 0.2561/T	0.56	-0.1732 + 71.3230/T	0.89	-0.3242 + 123.5891/T	1.43
CDFM +	1	0.0008 + 3.4123/T	1.63	-0.0045 – 13.2563/T	8.25	0.4856 – 169.9636/T	7.36
	4	0.0032 + 7.9452/T	1.63	-0.0082 – 10.6336/T	5.37	0.4215 – 138.3692/T	8.27
	6	-0.0166 + 12.2128/T	1.32	-0.0323 + 1.5215/T	5.74	0.1645 – 51.2363/T	6.83
TFM +	5	-0.0138 – 1.5262/T	0.16	-0.0189 + 18.3623/T	0.26	0.3125 – 89.2382/T	0.30
Global mean deviation			1.27		3.45		4.56

[a] $\Delta P = \frac{1}{NP}\sum_{i}^{NP}\frac{\left|P_{\exp,i} - P_{calc,i}\right|}{P_{\exp,i}}$

The single-phase region shrank and the slope of cloud point isopleth changed from a lower critical solution temperature (LCST) behavior to an upper critical solution temperature (UCST) behavior as the D,L-lactide content in PLAG copolymer decreased. The cloud point isopleths for the 65.27 : 34.73 and 49.60 : 50.40 proportions show an UCST curve, where the slope of the isopleths is negative in the P-T projection. The cloud point pressure deviations were 1.64, 2.69 and 4.46% for the PC-SAFT, SL and PR EoS, respectively.

5.5.2. PLAG + CO_2 System.

In Figure 5.5.2, the isopleths obtained by the three EoS for PLAG copolymer + CO_2 systems (the content of CO_2 in the binary mixture is 95.00 mass percent) are shown for two D,L-lactide : glycolide molar ratio, 65.00 : 35.00 (Mw/Mn = 1.43 and 1.76) and 85.00 : 15.00 (Mw/Mn = 1.54 and 1.73). All the isopleths have negative slope; some of them are steeper than others, but all seem to have a UCST-type phase behavior. Comparing the isopleths with the same molar ratio, for the ratio 65.00 : 35.00, it can be appreciated that, at constant temperature, as the polidispersity index increases, the cloud point pressure also increases.

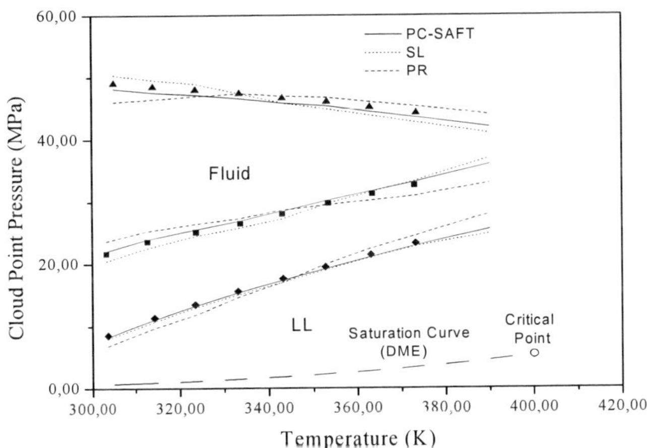

Figure 5.5.1. Cloud point isopleths for fluid-liquid equilibria of PLAG + DME systems (the content of DME in the binary systems is between 96.86 and 97.25 mass percents) at different mole ratios in poly(D,L-lactide-co-glycolide): ▲ = 65.27 : 34.73, ■ = 76.31 : 23.69, ◆ = 87.88 : 12.12. Experimental data were taken from Kuk et al. (2001).

The same thing occurs for the molar ratio 85.00 : 15.00 (Mw/Mn = 1.54 and 1.73). Another observation from this figure is that, when the polidispersity index remains constant, as the glycolide content in the PLAG copolymer increases, the cloud point pressure also increases; in other words, the one-phase region decreases. In this figure it can also be noticed the distance between the isopleths and the CO_2 saturation curve.

In Figure 5.5.3, the cloud point isopleths obtained with the PC-SAFT, SL and PR EoS are presented for PLAG copolymer + CO_2 systems with similar polidispersity indexes (1.73, 1.67 and 1.76) and with different D,L-lactide : glycolide molar ratios (85.00 : 15.00, 75.00 : 25.00 and 65.00 : 35.00), where the CO_2 has 95.00 mass percent in the binary mixture. As in Figure 5.5.2, here it can also be observed that the cloud point isopleths have a UCST-type phase behavior (negative slope) and that, when the glycolide content increases in the copolymer, the one-phase region decreases. According to Conway et al. (2001), the switch from a positive (PLA + CO_2 systems) to a negative slope (PLAG copolymers + CO_2 systems) suggests that the interchange energy, which is a measure of copolymer-CO_2, copolymer-copolymer and CO_2 - CO_2 interactions, is weighted more toward copolymer-copolymer interactions rather than cross-interactions. In others words, a cloud point curve with a negative slope shows that increasing the system pressure, or conversely, the solvent density, does not help in obtaining a single phase as the system temperature is lowered. Although CO_2 has a quadrupole moment of $-1.3598 \times 10^{-34}\ J^{1/2}m^{5/2}$, it is not expected that CO_2 forms any type of complex with PLAG. In terms of cloud point pressure deviations for the five PLAG copolymer + CO_2 systems, the results obtained with the PC-SAFT EoS (0.48%) are in good agreement with the experimental data, while the deviations obtained with the SL and PR EoS are slightly greater (1.38 and 1.24%, respectively).

5.5.3. PLAG + CDFM System

The solubilities of PLAG copolymers in CDFM were modeled based on literature data (Lee et al., 2000). The phase behavior of PLAG copolymer + CDFM systems for different molar ratio between D,L-lactide and glycolide in PLAG copolymer (70.00 : 30.00, 80.00 : 20.00 and 90.00 : 10.00), where the content of CDFM in these binary systems varies from 96.01 to 96.86 mass percents, is shown in Figure 5.5.4. For all the binary systems investigated, the LCST curve intersects the CDFM saturation curve and no U-LCST-type phase

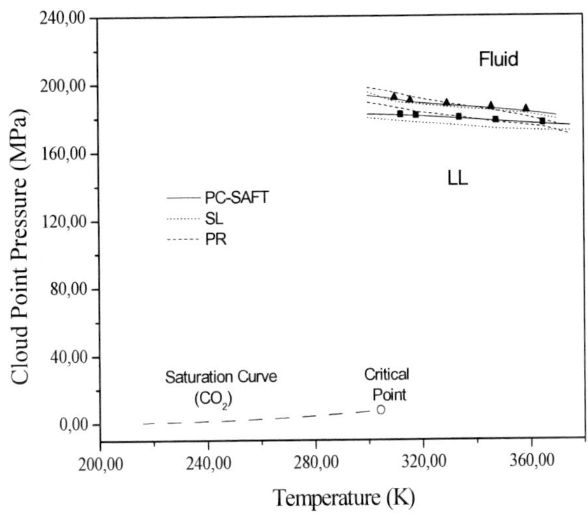

Figure 5.5.2. Cloud point isopleths for PLAG + CO_2 systems (the content of CO_2 in the binary systems is 95.00 mass percent) at two polidispersity indexes in PLAG copolymers: (D,L-lactide : glycolide: 85.00 : 15.00 in molar ratio) ■ = 1.54, ▲ = 1.73. Experimental data were taken from Conway et al. (2001).

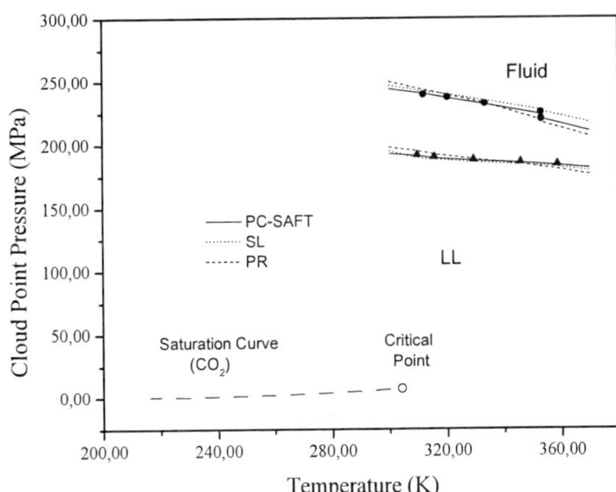

Figure 5.5.3. Cloud point isopleths for PLAG + CO_2 systems (the content of CO_2 in the binary systems is 95.00 mass percent) at two similar polidispersity indexes in PLAG copolymers (a) (D,L-lactide : glycolide: 85.00 : 15.00 in molar ratio) ▲ = 1.73, (b) (D,L-lactide : glycolide: 75.00 : 25.00 in molar ratio) ● = 1.67. Experimental data were taken from Conway et al. (2001).

behavior is detected. At similar mass percents of CDFM in PLAG copolymer + CDFM systems, when the D,L-lactide : glycolide molar ratio increases in glycolide, the one-phase region decreases. CDFM is a better solvent than other SCFs because its hydrogen atoms are probably more acidic than the ones in others SCFs (Conway et al., 2001). In terms of cloud point pressure deviations, which appear in Table 5.5.4 for the three PLAG copolymer + CDFM systems, the results obtained with the PC-SAFT EoS (1.40%) are in good agreement with the experimental data, while the other models can also correlate the phase behavior of PLAG copolymer + CDFM systems but with less precision than PC-SAFT EoS (5.63% for SL and 6.16% for PR).

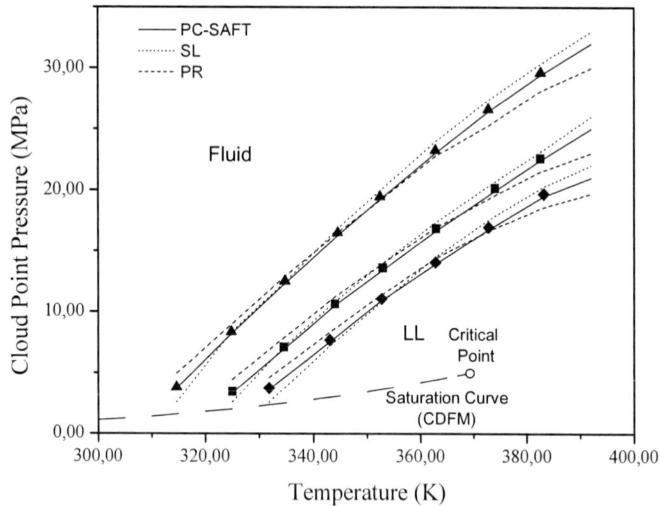

Figure 5.5.4. Cloud point isopleths for fluid-liquid equilibria of PLAG + CDFM systems (the content of CDFM in the binary systems is between 96.01 and 96.86 mass percents) at different mole ratios in poly(D,L-lactide-co-glycolide): ▲ = 70.00 : 30.00, ■ = 80.00 : 20.00, ◆ = 90.00 : 10.00. Experimental data were taken from Lee et al. (2000).

5.5.4. PLAG + TFM System

The phase behavior of PLAG copolymer in TFM is shown in Figure 5.5.5; the content of TFM in the binary system is 95.00 mass percent, the polidispersity index is 1.67 and the D,L-lactide : glycolide molar ratio is 75.00

: 25.00. In this figure, the cloud point pressures increase with temperature, indicating that this binary system exhibits the characteristics of a LCST phase behavior. The cloud point isopleths are far from the TFM saturation curve, and their intersection would happen probably at lower temperatures. Although CO_2 and TFM have approximately the same polarizability and both have some polarity, since TFM has a dipole moment of 1.649 D [DIPPR, 2000] and CO_2 has a quadrupole moment, it is expected that TFM forms a complex with PLAG because it has an acidic proton that is able of hydrogen bonding with the ester groups in PLAG. In this way, the ability of TFM to form a complex with PLAG makes it a better solvent than CO_2, especially since any change in favorable energetic interactions is magnified in these dense SCF solvents (Conway et al., 2001).

In Figure 5.5.5, the internal square, in extended scale and without the TFM saturation curve, shows the isopleths obtained with the three thermodynamic models, which are also presented in the external square, for a better visualization of the results of the phase behavior predicted for each EoS. The cloud point pressure deviations are all of the same order of magnitude, 0.16% for PC-SAFT, 0.26% for SL and 0.30% for PR, indicating that all models are able to represent this system.

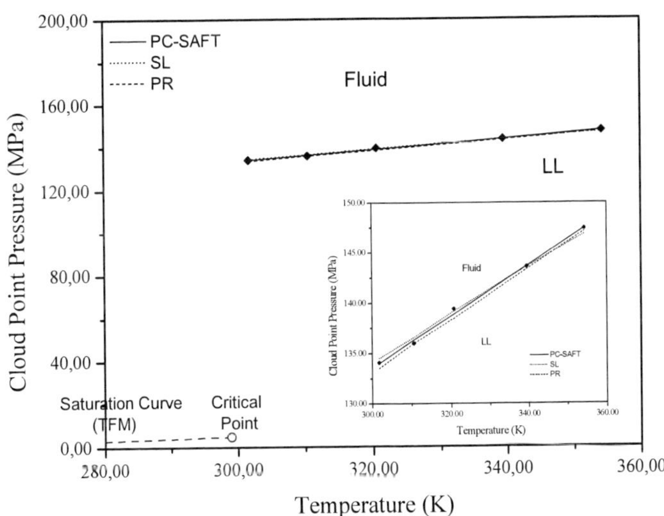

Figure 5.5.5. Cloud point isopleths for PLAG + TFM systems (the content of TFM in the binary systems is 95.00 mass percent) at one polidispersity index in PLAG copolymers (D,L-lactide : glycolide: 75.00 : 25.00 in molar ratio) ◆ = 1.76. Experimental data were taken from Conway et al. (2001).

5.6. Block Copolymer + Supercritical CO_2 Systems

In this section, the high-pressure phase behavior of binary systems diblock copolymer + supercritical fluid is studied. Table 5.6.1 presents the PC-SAFT parameters of the polymers for which PVT data are found in the literature; these parameters are obtained was explained in section 5.1.2.

Table 5.6.1. PC-SAFT pure-component parameters for polymers whit available PVT data.[1]

Polymer	m/PM $(10^{-3}$ kg/mol$)^{-1}$	σ (m × 10^{10})	ε/k (K)	$\Delta P^{L\ a}$	$\Delta v^{L\ b}$
p(EO)	0.080017	2.5125	230.12	0.2156	0.4743
p(VAc)	0.029905	3.5086	310.14	0.0326	0.3265

$^{a}\ \Delta P^{L} = \frac{1}{NP}\sum_{i}^{NP}\frac{\left|P^{L}_{\exp,i} - P^{L}_{calc,i}\right|}{P^{L}_{\exp,i}}$, $^{b}\ \Delta v^{L} = \frac{1}{NP}\sum_{i}^{NP}\frac{\left|v^{L}_{\exp,i} - v^{L}_{calc,i}\right|}{v^{L}_{\exp,i}}$

The polymers for which no experimental PVT data are avaliable in literature are shown in Table 5.6.2. In this table there are also shown the PHSC pure-component parameters, which were calculated using the Elvassori et al. (2002) method, explained in section 5.1.2. By using these PHSC parameters, the PVT data were generated and used for estimating the PC-SAFT parameters, which are shown in Table 5.6.3.

Table 5.6.2. Some physical characteristics and PHSC pure-component parameters for polymers whith no PVT data available.[2]

Polymer	T (K)	Polidispersivit y	PHSC pure-component parameters		
			A* 10^8 (m^2/mol)	V* 10^3 (m^3/mol)	E* (MPa m^3/ mol)
p(PO)	300 - 500	1.80	8.1465	28.1046	6.1520
p(BO)	300 - 450	0.93	6.2585	23.1083	8.8128
p(AHO)	300 - 450	1.20	8.1404	29.4314	8.4043

[1] p(EO): poly(ethylene oxide), p(VAc): poly(vinyl acetate).
[2] p(PO): poly(propylene oxide), p(BO): poly(butylene oxide), p(AHO): poly(1,1,2,2 tetrahydroperfluorooctyl acrylate).

Table 5.6.3. PC-SAFT pure-component parameters for polymers listed in Table 5.6.2

Polymer	m/MW $(10^{-3}$ kg/mol$)^{-1}$	σ (m × 10^{10})	ε/k (K)	$\Delta P^{L\ a}$	$\Delta v^{L\ b}$
p(PO)	0.028145	3.8218	314.45	0.0148	0.2815
p(BO)	0.026133	3.9314	325.04	0.0124	0.3210
p(AHO)	0.021043	2.2815	301.43	0.0293	0.4146

$^{a}\ \Delta P^{L} = \frac{1}{NP}\sum_{i}^{NP}\frac{\left|P_{\exp,i}^{L} - P_{calc,i}^{L}\right|}{P_{\exp,i}^{L}}$, $^{b}\ \Delta v^{L} = \frac{1}{NP}\sum_{i}^{NP}\frac{\left|v_{\exp,i}^{L} - v_{calc,i}^{L}\right|}{v_{\exp,i}^{L}}$

Table 5.6.4 presents some physical characteristics of the binary systems, while Table 5.6.5 shows the properties and the copolymerization parameters (segment fraction and bonding fraction) obtained using the PC-SAFT EoS for the diblock copolymer + supercritical CO_2 systems. It is noticed that the bonding fraction parameters correspond for the alternative arrangement, such as it was explained in Table 3.1.

Table 5.6.4. Main characteristics of binary systems diblock copolymer + supercritical CO_2.

Diblock copolymer	Copolymer composition	Supercritical fluid	$MW_{\text{block copolymer}}$ (g/mol)	Ref
p(EO-b-BO)	44.40% mole EO	CO_2	1 620	[a]
p(EO-b-PO)	8.60% mole EO		1 988	
	10.00% mole EO		566	
p(VAc-b-AHO)	50.00% mole VAc		70 700	[b]

[a] O'Neill et al. (1998), [b] Colina et al. (2002).

Table 5.6.5. Properties and PC-SAFT copolimerization parameters of copolymers for the phase equilibrium modeling of diblock copolymer + supercritical CO_2 systems

SCF	Diblock copolymer	Diblock copolymer composition			Segment fraction	Bonding fraction		
		mole %	mass %	segment β	z_β	$B_{\alpha\alpha}$	$B_{\alpha\beta}$	$B_{\beta\beta}$
CO_2	p(EO-b-BO)	44.40	32.79	EO	0.5000	1.0000	0.0000	0.0000
	p(EO-b-PO)	8.60	6.66	EO	0.5000	1.0000	0.0000	0.0000
		10.00	7.77					
	p(VAc-b-AHO)	50.00	14.37	VAc	0.5000	1.0000	0.0000	0.0000

In Table 5.6.6 there are shown the binary interaction parameters correspondents to the segment-segment interactions and the pressure deviations in the cloud point of diblock copolymer + supercritical CO_2 systems. From these deviations, it is evident the good performance of PC-SAFT EoS for modeling the binary systems formed by block copolymers and supercritical fluids.

Table 5.6.6. Binary interaction parameter and pressure deviations obtained with PC-SAFT EoS for modeling the phase equilibra of block copolymer + supercritical CO_2 systems

SCF	Diblock copolymer	Segment-segment pair	$\kappa_{i\alpha j\beta}$	ΔP [a]
CO_2	p(EO-b-BO)	CO_2 + p(EO)	0.0148	1.43
		CO_2 + p(BO)	0.0183	
		p(EO) + p(BO)	0.0104	
	p(EO-b-PO)	CO_2 + p(EO)	0.0148	1.38
		CO_2 + p(PO)	0.0215	
		p(EO) + p(PO)	0.0089	
	p(AV-b-AHO)	CO_2 + p(AV)	0.0149	1.59
		CO_2 + p(AHO)	0.0025	
		p(AV) + p(AHO)	0.0104	

[a] $\Delta P = \frac{1}{NP}\sum_{i}^{NP}\frac{\left|P_{\exp,i} - P_{calc,i}\right|}{P_{\exp,i}} * 100$

5.6.1. p(EO-b-BO) + CO_2 System

In Figure 5.6.1, the phase equilibrium of p(EO) + CO_2 systems is shown at several concentrations. For these systems, which exhibit a LCST-type phase behavior, the solubility decreases when temperature increases at constant pressure. That tendency is due to the decrease of the solvent density with the temperature. Density, and not pressure, is the natural variable to understand the supercritical fluid solvatation. Density describes the free volume, which has an important influence in the phase separation of LCST type. When the solvent density increases, its free volume also increases, resulting in an increment of the interactions between the polymer-polymer segments due to the located densification. That provokes the expansion of the solvent to a new phase, increasing the mixture volume and entropy. Like this, p(EO) + CO_2

systems have an increase of the cloud pressures with the temperature for a certain concentration of polymer, indicating a LCST-type phase behavior. On the other hand, p(BO) has one more ethyl group (-C_2H_5) than p(EO), which increases significantly the p(BO) solubility when compared to the p(EO) solubility in CO_2; however, p(BO) is less soluble than p(PO), which is demonstrated through their high cloud point pressures.

Probably this decrease of the solubility is partly due to the fact that the substitutions add a great increment in superficial tension that a substitution of a methyl group (-CH_3), as it is the case of p(PO). Esteric factors can also be the cause of the decrease of the p(BO) solubilities when compared with p(PO). Since it is known that the phase behavior of p(EO) + CO_2 and p(BO) + CO_2 systems is of LCST-type, then it is easy to deduce that the phase behavior of p(EO-b-PO) + CO_2 systems will be also of LCST-type. That is confirmed in Figure 5.6.2 for the p(EO-b-BO) copolymer (44% mole EO in the copolymer) for two temperatures (296.15 and 308.15 K). PC-SAFT EoS was be able to predict with success the LCST-type phase behavior of this binary system.

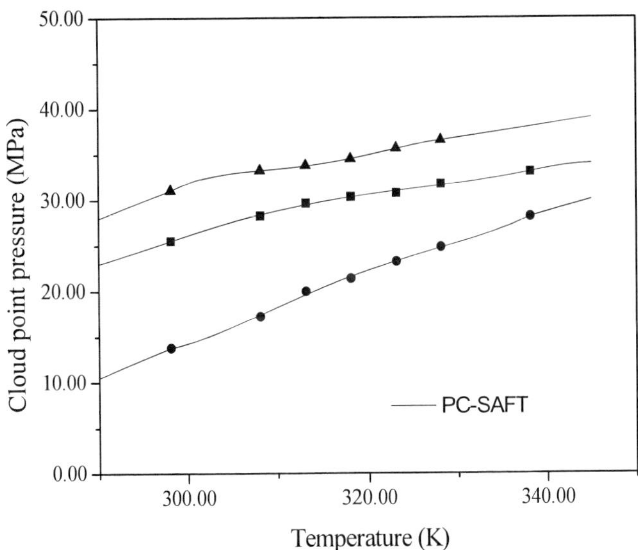

Figure 5.6.1. Cloud point isopleths for p(EO) (MW = 600) + CO_2 systems at several polymer compositions. Experimental data (● = 0.296% mass, ■ = 0.610% mass, ▲ = 0.924% mass polymer) were taken from O'Neill et al. (1998).

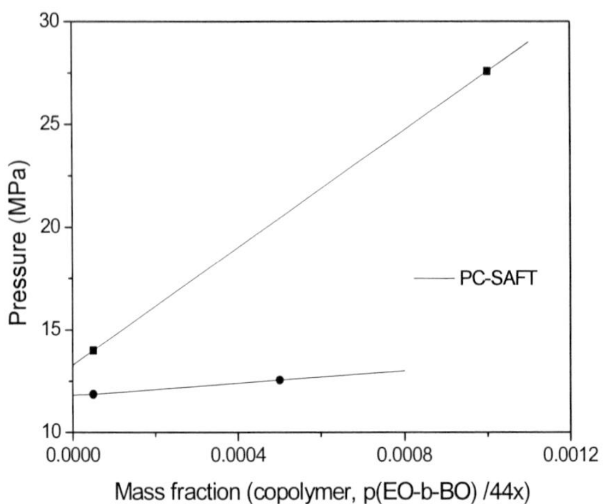

Figure 5.6.2. Pressure vs. copolymer mass fraction projection of p(EO-b-BO) + supercritical CO_2 systems. Experimental data (● = 296.15 K, ■ = 308.15 K) were taken from O'Neill et al. (1998).

5.6.2. p(EO-b-PO) + CO_2 System

Experimental (O'Neill et al., 1998) and calculated data of the phase behavior of p(PO) polymers (MW = 400) + CO_2 systems are shown in Figure 5.6.3. The polymer p(PO) is completely soluble in CO_2 above 10% in mass at 303.15 K, and 9.03 MPa. The polymer p(PO) is much more soluble than p(EO) for a certain molecular weight. It is clearly noticed that the substitution of a methyl group in each monomer in the chains has a great effect on solubility, but the LCST-type phase behavior is always maintained. Summarizing, the greater solubility of p(PO) compared to p(EO) in CO_2 is due to the weak segment-segment interactions, as well as to the weak self-associations between the segments of p(EO) in CO_2. Binary systems p(EO-b-PO) + CO_2 also present the LCST-type transition phase behavior, due to that the polymers which form the diblock copolymer [p(EO) and p(PO)], also present the same phase behavior when CO_2 is present (Figures 5.6.1. and 5.6.3.). The only difference with the previous system [p(EO-b-PO) + CO_2] is that this diblock copolymer [p(EO-b-PO)] is more soluble in CO_2 (Figure 5.6.4) when the pressures are compared with the cloud points in Figure 5.6.2. Although the diblock copolymers p(EO-b-PO) present different molecular weights (1 988 and 566), when mixed with CO_2 and at phase equilibrium, they present the LCST-type phase behavior.

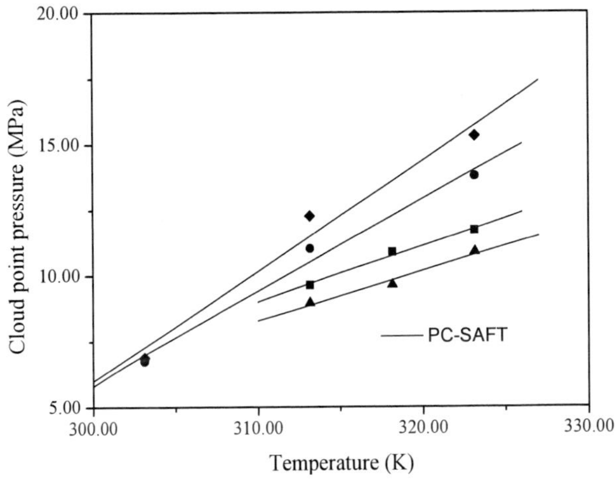

Figure 5.6.3. Cloud point isopleths for p(PO) (MW = 400) + CO_2 systems at several polymer compositions. Experimental data (▲= 0.35% mass, ■= 1.00% mass, ● = 3.10% mass, ◆ =9.50% mass polymer) were taken from O'Neill et al. (1998).

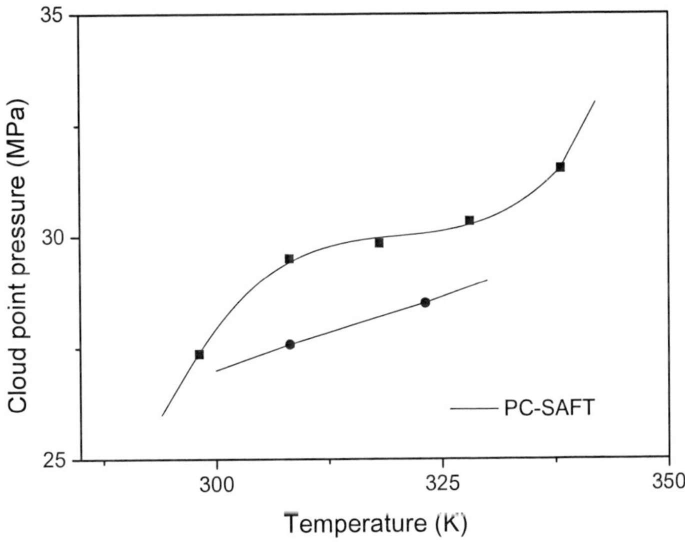

Figure 5.6.4. Cloud point isopleths of p(EO-b-PO) + supercritical CO_2 systems. Experimental data (●= 8.60% mass, ■= 10.00% mass copolymer) were taken from O'Neill et al. (1998).

5.6.3. P(Vac-B-AHO) + CO_2 System

In this section, the ability of the PC-SAFT EoS to describe the cloud point curves (liquid-liquid transition) for the p(VAc) + CO_2 and p(AHO) + CO_2 system will be discussed. After that, the modeling of the cloud point curves of the p(VAc-b-AHO) + CO_2 system will be commented on. In Figure 5.6.5 there are shown the chemical structures of the p(VAc) and p(AHO) polymers.

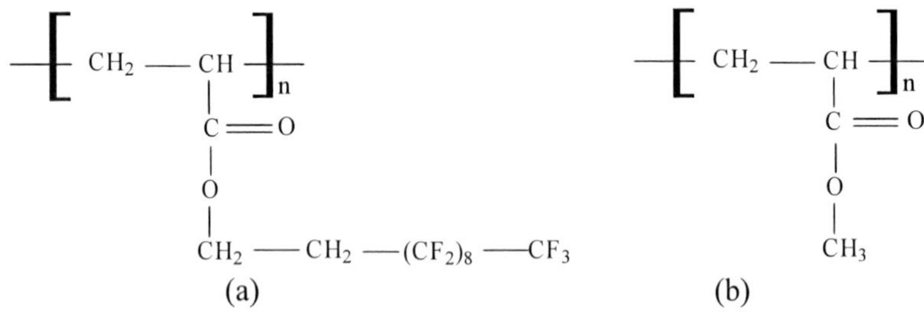

Figure 5.6.5. Chemical structure for polymers: (a) p(AHO), (b) p(VAc).

Mawson et al. (1995) obtained experimental data of the cloud point curves of p(AHO) polymer (MW = 60 400 g/mole) in CO_2 at several concentrations, varying from 0.087% to 7.32% mass. The cloud point curves of p(AHO) + CO_2 system presents a LCST-type behavior, as shown in Figures 5.6.6 and 5.6.7. This type of phase separation is typical for polymer mixtures in supercritical fluids. The p(VAc) + CO_2 system presents a similar behavior, as demonstrated by Buhler et al. (1998). Both systems were modeled with good precision by the PC-SAFT EoS when the cloud point isotherms were compared at experimental conditions with the predicted curves.

In Figure 5.6.8, the chemical potentials of CO_2 predicted by the PC-SAFT EoS and those predicted by Colina et al. (2002) are shown against the concentration of p(VAc-b-AHO) at 45°C and 13.5 and 44.0 MPa, using the parameters presented in Tables 5.6.5 and 5.6.6.

It should be noticed that the curves present different behavior with the variation of pressure. At 13.5 MPa, the chemical potential is almost independent on the copolymer concentration at low concentrations, then appears an inflection point and, finally, at high concentrations, it remains constant. At 44.0 MPa, the chemical potential varies almost linearly with the copolymer concentration.

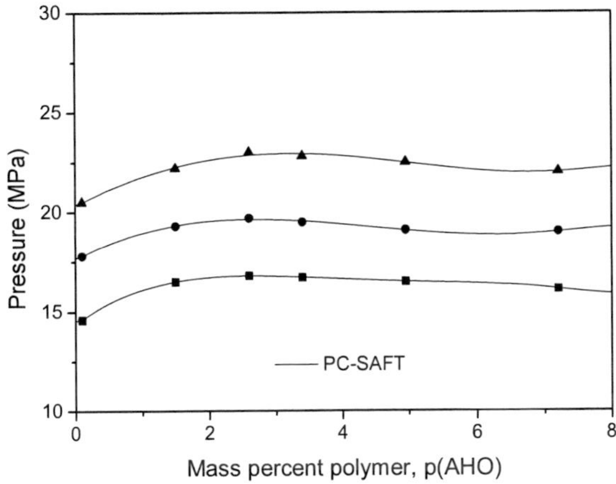

Figure 5.6.6. Cloud point isotherms for p(AHO) + CO_2 systems. Experimental data (■ = 313.15 K, ● = 323.15 K, ▲ = 333.15 K) were taken from Mawson et al. (1995).

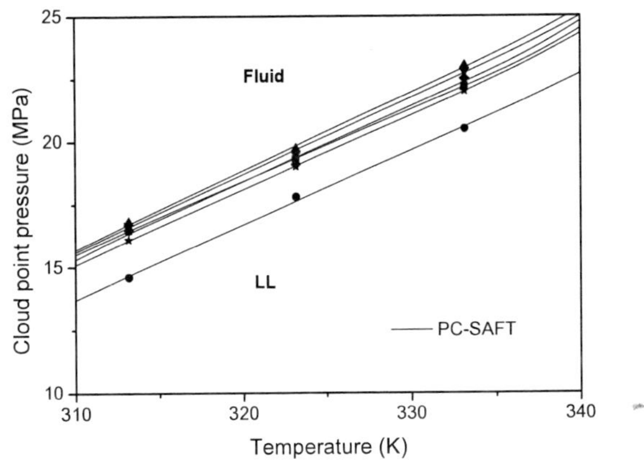

Figure 5.6.7. Cloud point isopleths of p(AHO) + supercritical CO_2 systems. Experimental data (● = 0.087% mass, ■ = 1.502% mass, ▲ =2.603% mass, ▼ = 3.408% mass, ◆ = 4.943% mass, ж = 7.320% mass polymer) were taken from Mawson et al. (1995).

In Figure 5.6.9, the phase diagram of p(VAc-b-AHO) in CO_2 at 318.15 K is shown. In this figure, there are defined three regions: a two-phase (LL) region, a one-phase (fluid phase) region and, between these, there is a metaestable region. In the interface between the two-phase and the metaestable regions, the CO_2 density is low (less than 0.82 g/cm^3). Figure 5.6.9 indicates that PC-SAFT EoS seems to be capable to model the phase equilibrium of p(VAc-b-AHO) + CO_2 systems.

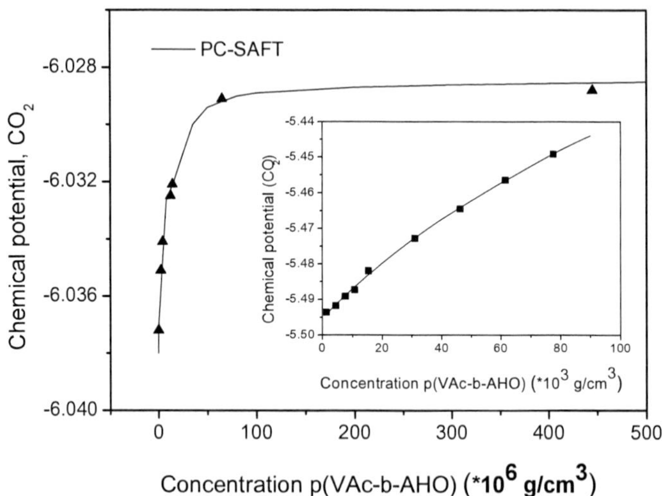

Figure 5.6.8. CO_2 chemical potentials vs. copolymer concentration projection of p(VAc-b-AHO) + supercritical CO2 systems. Data points (▲ = 13.5 MPa and 318.15 K, ■ = 44.0 MPa and 318.15 K) predicted by Colina et al. (2002).

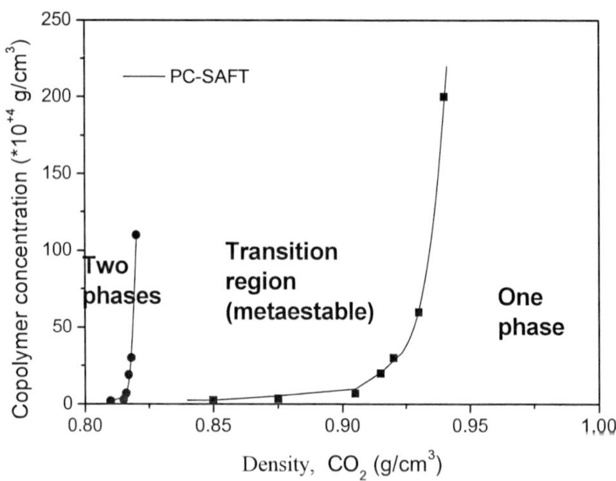

Figure 5.6.9. Phase diagram for p(VAc-b-AHO) in supercritical CO_2 ($MW_{p(AHO)}$ = 60 400 g/mole, $MW_{p(VAc)}$ = 10 300 g/mole). Experimental data (● = Limit between the two-phase and metaestable region, ■ = limit between the metaestable and one-phase region) were taken from Buhler et al. (1998).

5.7. Polymer Blends and Polymer Blend + CO_2 Systems

In this section, the PC-SAFT and SL EoS were used to correlate and predict the thermodynamic behavior of the LLE of polymer blends (PBD/PS, PPG/PEGE, PVME/PS, and PEO/PES)[3], as well as the fluid phase behavior of PVME/PS blends at various compositions in the presence of CO_2. PC-SAFT and SL pure-component parameters for each polymer were obtained by fitting liquid pure-component PVT data (Rodgers, 1993) over a pressure range suitable for engineering calculations, such as it was explained in section 5.1.2. But for the PEGE, PPG and PES there are no PVT data available in literature. Then, by applying the Elvassore et al. (2002) method, PVT data are generated. In Table 5.7.1, some physical characteristics for these polymers and their PHSC pure-component parameters are shown, while Tables 5.7.2a and 5.7.2b show the PC-SAFT and SL pure-component parameters, respectively, calculated as explained in section 5.1.2.

Table 5.7.1 Polymer properties and PHSC pure-component parameters

Polymer	T (K)	Polidispersity (Mw/Mn)	PHSC pure-component parameters		
			A^* 10^8 (m^2/mol)	V^* 10^3 (m^3/mol)	E^* (MPa m^3/mol)
PEGE	300 - 400	1.03	7.8526	23.4120	5.4523
PPG	300 - 400	[a]	6.1452	25.7415	5.8542
PES	320 - 450	1.05	6.2563	24.5412	6.1524

[a]Polidispersitiy index is not available. In this case, it was used $Mw/Mn = 1.00$.

Table 5.7.2a. PC-SAFT pure-component parameters for polymers

Polymer	PC-SAFT EoS				
	m/MW (10^{-3} kg/mol)$^{-1}$	σ (m × 10^{10})	ε/k (K)	$\Delta P^{L\ a}$	$\Delta v^{L\ b}$
PBD	0.060552	2.9845	240.2132	0.06	0.30
PEGE	0.043561	3.1526	290.5413	0.08	0.39

[3] PBD: poly(butadiene), PEGE: poly(ethylene glycol mono-methyl ether), PPG: poly(ethylene glycol), PS: polystyrene, PVME: poly(vinyl methyl ether), PEO: poly(ethylene oxide), PES: poly(ether sulfone).

PEO	0.080017	2.5125	230.1245	0.22	0.47
PES	0.023433	3.6148	356.1422	0.10	0.31
PPG	0.040123	3.3342	310.5026	0.07	0.46
PS	0.033238	3.5022	320.1372	0.02	0.29
PVME	0.024245	3,9145	372.4584	0.18	0.24

[a] $\Delta P^L = \frac{1}{NP}\sum_i^{NP}\frac{\left|P^L_{\exp,i} - P^L_{calc,i}\right|}{P^L_{\exp,i}}$, [b] $\Delta v^L = \frac{1}{NP}\sum_i^{NP}\frac{\left|v^L_{\exp,i} - v^L_{calc,i}\right|}{v^L_{\exp,i}}$

Table 5.7.2b. SL pure-component parameters for polymers

Polymer	SL EoS				
	T* (K)	P* (MPa)	ρ* (kg/m³)	ΔP^{L} [a]	Δv^{L} [b]
PBD	562.13	412.41	972.41	1.54	1.80
PEGE	630.21	502.32	1020.12	2.05	1.75
PEO	663.40	487.44	1168.23	1.86	2.13
PES	881.14	645.10	1389.18	0.26	0.22
PPG	615.12	520.41	11140.63	1.23	1.12
PS	731.25	361.23	1112.36	0.57	1.22
PVME	580.52	392.45	1082.41	0.62	1.16

[a] $\Delta P^L = \frac{1}{NP}\sum_i^{NP}\frac{\left|P^L_{\exp,i} - P^L_{calc,i}\right|}{P^L_{\exp,i}}$, [b] $\Delta v^L = \frac{1}{NP}\sum_i^{NP}\frac{\left|v^L_{\exp,i} - v^L_{calc,i}\right|}{v^L_{\exp,i}}$

5.7.1. Cloud Points Temperature of Blends

Table 5.7.3 lists the results obtained for the polymer blend miscibility (cloud point temperature) for four blends with different polymers. In this table also appear the binary interaction parameters and the cloud point temperature deviations. It can be seen that the binary interaction parameters depend on the molecular weights of the two polymers which form the blend. From the temperature deviations it is possible to deduce that both thermodynamic models, PC-SAFT and SL EoS, are able to correlate the fluid phase behavior of these polymer blend systems with reasonable accuracy. A detailed discussion appears in the subsections below.

Table 5.7.3. Results of the cloud point temperatures for polymer blend systems

Blends	MW $_{pol\,1}$/MW $_{pol\,2}$	PC-SAFT EoS		SL EoS	
		κ_{ij}	ΔT [a]	κ_{ij}	ΔT [a]
PBD/PS	1 100 / 1 340	0.0123	1.03	0.0645	3.43
	1 100 / 1 670	0.0112		0.1087	
	1 100 / 4 370	0.0140		0.0840	
	2 350 / 1 900	0.0095		0.1021	
	2 350 / 3 300	0.0125		0.0963	
PVME/PS	99 000 / 50 000	0.0094	1.11	0.1143	2.81
	99 000 / 100 000	0.0081		0.0815	
	95 000 / 67 000	0.0120		0.1016	
	95 000 / 106 000	0.0106		0.0915	
PPG/PEGE	2 000 / 550	0.0083	1.14	0.1028	3.04
	2000 / 750	0.0093		0.0643	
PEO/PES	20 000 / 4 000	0.0103	0.98	0.0443	2.59
	20 000 / 20 000	0.0126		0.0506	
	20 000 / 200 000	0.0096		0.0613	

[a] $\Delta T = \frac{1}{NP}\sum_{i}^{NP}\frac{\left|T_{\exp,i} - T_{calc,i}\right|}{T_{\exp,i}} * 100$

5.7.1.1. PBD/PS Blends

Figure 5.7.1 shows the phase behavior diagrams of PBD/PS systems with different molecular weights corresponding to experimental data reported by Rostami and Walsh (1985) and the results obtained with the PC-SAFT and SL models. In this figure it is observed a UCST-type phase behavior for all molecular weights. The temperature deviations are 1.03% for PC-SAFT and 3.43% for SL EoS. Thus, both models describe successfully these polymer blends on the whole concentration and temperature range. In all cases, for these polymer blends, when the molecular weight of the PBD segments remains constant and the molecular weight of the PS segments increases, the value of UCST tends to increase; in others words, the UCST value depends directly on the molecular weight of one of the polymer segments which forms the blend, in this particular case the polymer PS.

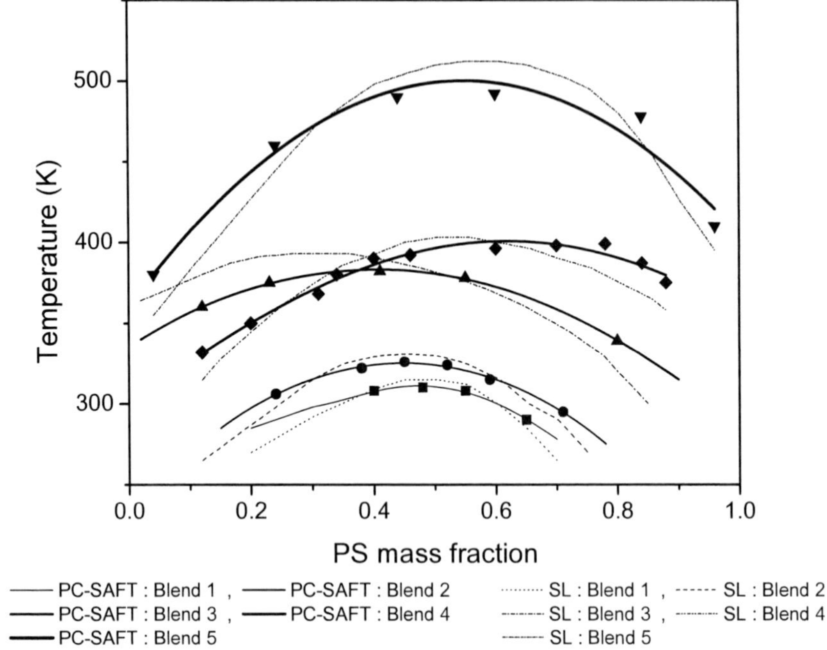

Figure 5.7.1. PC-SAFT and SL correlation of cloud point temperature behavior of PBD/PS blends. Experimental data (■: Blend 1, [PBD (1 100)/PS (1 340)]; ●: Blend 2, [PBD (1 100)/PS (1 670)]; ▲: Blend 3, [PBD (1 100)/PS (4 370)]; ◆: Blend 4, [PBD (2 350)/PS (1 900)]; ▼: Blend 5, [PBD (2 350)/PS (3 300)]) were taken from Rostami and Walsh (1985).

5.7.1.2. PVME/PS Blends

Figure 5.7.2 shows the polymer blend miscibility diagrams of PVME/PS systems corresponding to the experimental data presented by Bae et al. (1993) and Walsh et al. (1989) and the phase diagrams obtained with the PC-SAFT and SL EoS. These systems have an oriented interaction between the PVME chains and exhibit a LCST behavior. In this figure it can be noticed that, for blends 2 and 4, if PVME molecular weight remains constant (99 000) and PS molecular weight is 50 000 (blend 4), the value of its LCST is higher than the blend 2, when PS molecular weight is 100 000. A similar behavior is observed when there are compared the LCSTs of blends 1 and 3. The value of the LCST is higher when the PS molecular weight is 67 000 than when is 106 000 in blend 1. Both thermodynamic models managed to correlate and predict with high precision the LLE of PVME/PS blends, as well as the LCST; the PC-

SAFT EoS is more efficient in terms of the cloud point temperature deviations (1.11% for PC-SAFT and 2.81% for SL).

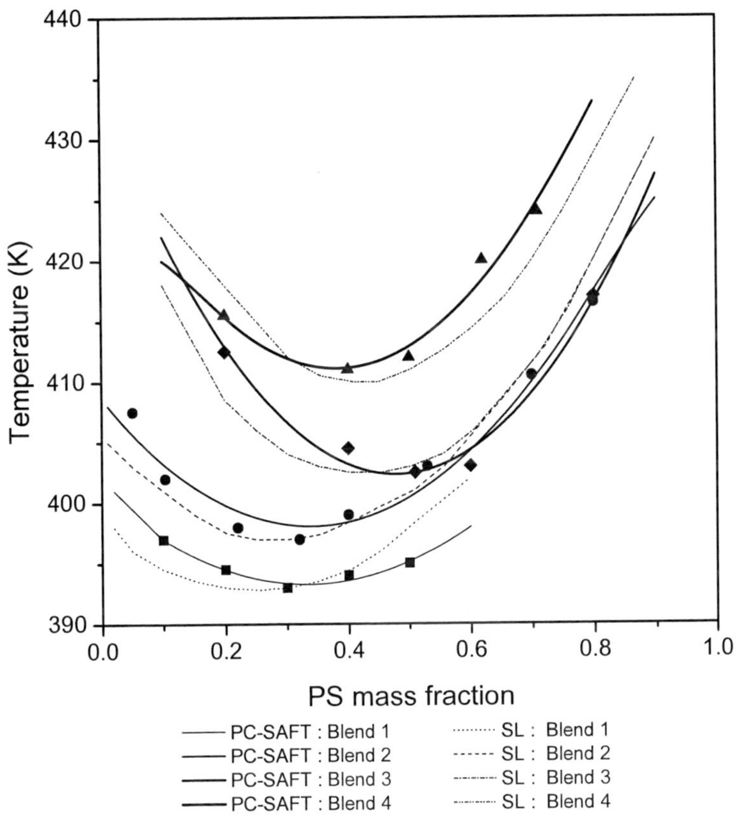

Figure 5.7.2. PC-SAFT and SL correlations and predictions of cloud point temperature behavior of PVME/PS blends. Experimental data (■: Blend 1, [PVME (95 000)/PS (106 000)]; ●: Blend 2, [PVME (99 000)/PS (100 000)]; ◆: Blend 3, [PVME (95 000)/PS (67 000)]; ▲: Blend 4, [PVME (99 000)/PS (50 000)]) were taken from Bae et al. (1993) and Walsh et al. (1989).

5.7.1.3. PPG/PEGE Blends

Figure 5.7.3 shows the experimental data reported by Takahashi et al. (1991) for PPG/PEGE systems and those obtained with the PC-SAFT EoS. In these blends, there are two different polymer chains which interact strongly, so there is a proper orientation to each other. In the same way that for the PBD/PS blends, the PPG/PEGE blends present also the UCST-type phase behavior. In this figure there are shown two blends; both of them have PPG segments which constant molecular weight (2 000) and PEGE segments (550 and 750 for both blends). The blend which contains PEGE segments with high molecular weights presents a higher UCST value than other PPG/PEGE blends with lower molecular weight of PEGE segments. The cloud point temperature deviations obtained by the PC-SAFT and SL models were 1.14% and 3.04%, respectively.

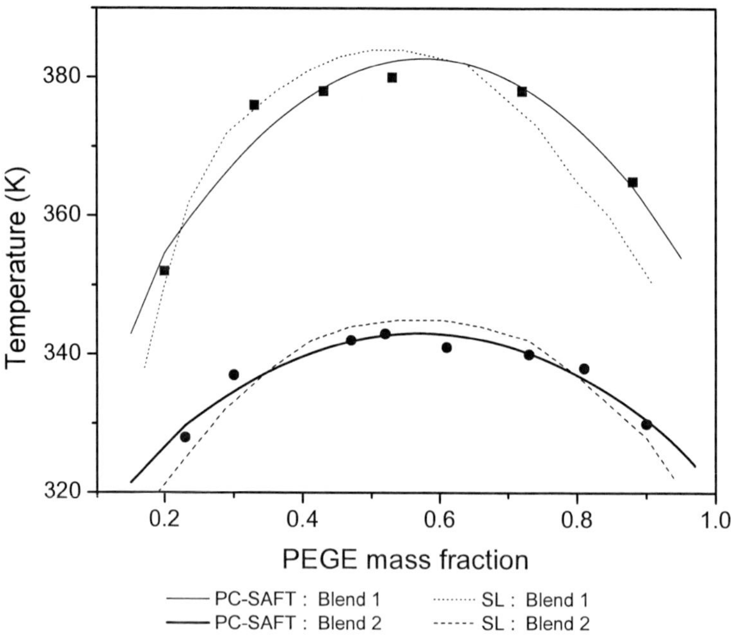

Figure 5.7.3. PC-SAFT and SL correlations and predictions of cloud point temperature behavior of PPG/PEGE blends. Experimental data (■: Blend 1, [PPG (2 000)/PEGE (750)]; ●: Blend 2, [PPG (2 000)/PEGE (550)]) were taken from Takahashi et al. (1991).

5.7.1.4. PEO/PES Blends

In Figure 5.7.4, the miscibility behavior of three PEO/PES blend predicted by the PC-SAFT and SL models are shown, together with experimental data by Walsh (1986. These PEO/PES blends are completely miscible at low temperature due to specific interaction between the polymer segments, but partially miscible at higher temperatures, where these interactions weaken and free-volume effects dominate (Walsh, 1986; Voutsas et al., 2004). From this figure it is also possible to observe the LCST-type phase behavior for the three PEO/PES blends, differentiated only by the change of the molecular weight of PEO segments, while the molecular weight of PES segments remains constant. The effect of the molecular weight of the segments of one of the polymers on the LCST of the blend is also apparent. While the molecular weight of PEO segments (4 000 - 200 000) increases, the value of the LCST decreases, maintaining constant the molecular weight of PES segments (20 000). Even though both thermodynamic models managed to correlate in a reasonable way the LLE of PEO/PES blends, the PC-SAFT EoS has a higher accuracy for represent the polymer blend miscibilities in terms of the cloud point temperature deviations than the SL EoS (0.98% and 2.59%, respectively).

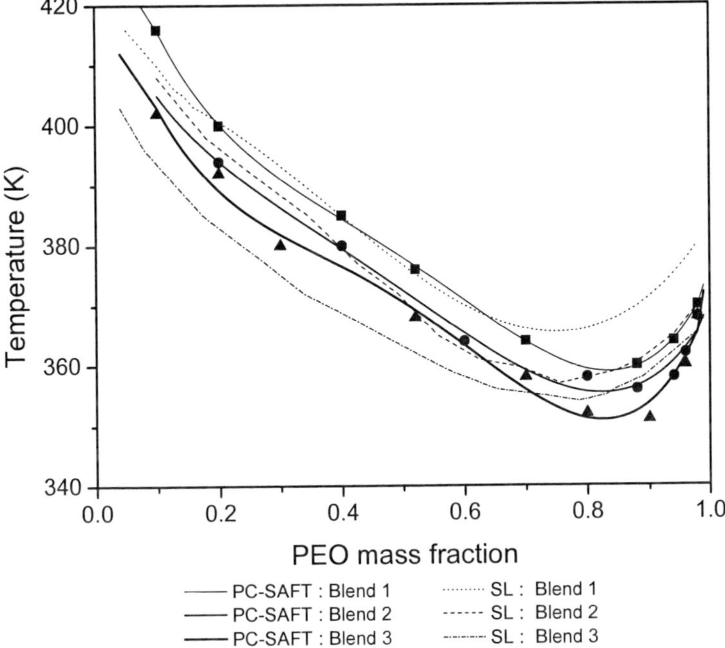

Figure 5.7.4. PC-SAFT and SL correlations and predictions of cloud point temperature behavior of PEO/PES blends. Experimental data (■: Blend 1, [PEO (4 000)/PES (20 000)]; ●: Blend 2, [PEO (20 000)/PES (20 000)]; ▲: Blend 3, [PEO (200 000)/PES (20 000)]) were taken from Walsh (1986).

5.7.2. Fluid Phase Behavior of PS/PVME Blend + CO_2 Systems

Since each of the three binary interaction parameters is known independently (Table 5.7.4), the calculation can in principle be conducted without any additional parameters. It is instructive to first assess how the EoS formalism performs for predicting the solubility of CO_2 in the pseudo binary PS/PVME + CO_2 system. In Figure 5.7.5, the results and experimental data from Mokdad et al., 1996 for the CO_2 sorption isotherms for PVME/PS (50/50) blend + CO_2 at 293.15 and 313.15 K are shown. The results indicate a good agreement with the experimental data. As shown in Figures 5.7.6a and 5.7.6b, the experimental sorption isotherms for CO_2 are linear with the equilibrium pressure. In these figures, it is easy to notice that CO_2 solubility increases with PVME concentration. Perhaps this can be explained by the effect of polarity on the PVME segment, which is higher than that of the PS one, so the dipole-dipole interactions with CO_2 increase with the concentration in PVME. From these figures it can also be observed that the isotherms for higher PS concentrations show a slight concavity towards the pressure axis.

Table 5.7.4. Pressure deviations and binary interaction parameters for CO_2 + polymer, blends and blend + CO_2 systems

EoS	Temperature (K)		Systems [a]			
			CO_2 + PS [b]	CO_2 + PVME [c]	PVME/PS (50/50) [d]	ΔP^e PVME/PS (50/50) + CO_2
PC-SAFT	293.15	κ_{ij}	0.0634	0.0080	0.2047	2.58
		ΔP [e]	1.02	1.85	1.60	
	313.15	κ_{ij}	0.0603	0.0066	0.2059	2.89
		ΔP [e]	1.06	1.36	1.18	
SL	293.15	κ_{ij}	-0.0048	0.0254	0.4821	5.12
		ΔP [e]	3.05	3.40	3.85	
	313.15	κ_{ij}	-0.0029	0.0285	0.5041	4.89
		ΔP [e]	2.99	4.02	4.03	

[a]MW $_{PS}$ = 100 000; MW $_{PVME}$ = 99 000; [b]Arce (2005); [c]Experimental data by Wang (2006); [d]Arce and Aznar (2009); [e] $\Delta P = \frac{1}{NP}\sum_{i}^{NP}\frac{|P_{\exp,i} - P_{calc,i}|}{P_{\exp,i}}$

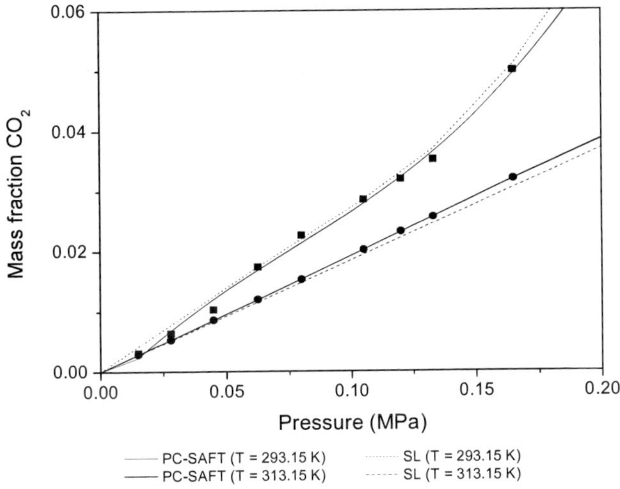

Figure 5.7.5. CO_2 sorption isotherms using the PC-SAFT and SL EoS in PS/PVME blends. Experimental data (■ = 293.15 K; ● = 313.15 K) were taken from Rao and Watkins (2000).

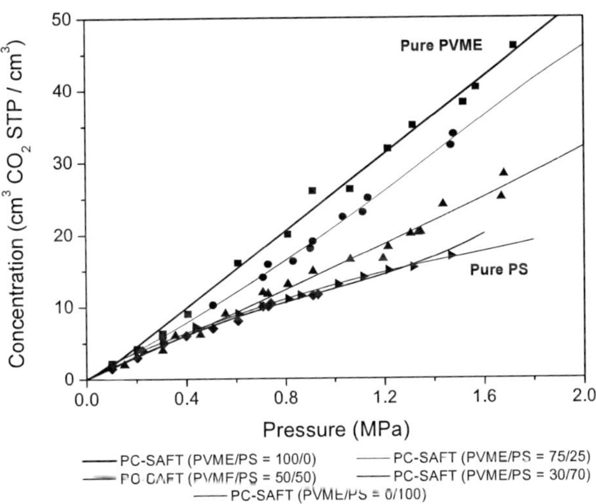

Figure 5.7.6a. CO_2 sorption isotherms obtained with the PC-SAFT EoS in PS/PVME blends (T = 293.15 K) at different blend compositions. Experimental data (■ = PVME/PS blend (100/0); ● = PVME/PS blend (75/25); ▲ = PVME/PS blend (50/50); ◆ = PVME/PS blend (30/70); ▶ = PVME/PS blend (0/100)) were taken from Mokdad et al. (1996).

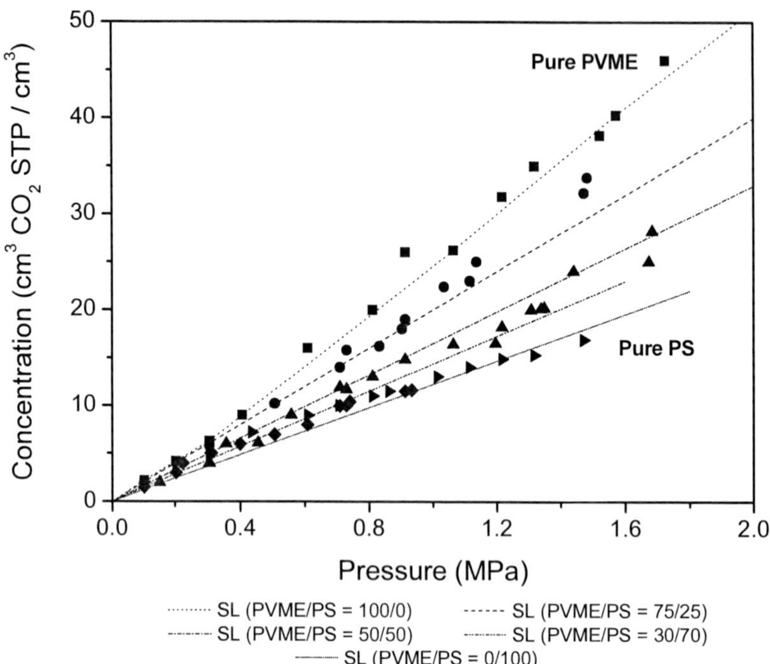

Figure 5.7.6b. CO_2 sorption isotherms obtained with the SL EoS in PS/PVME blends (T = 293.15 K) at different blend compositions. Experimental data (■ = PVME/PS blend (100/0); ● = PVME/PS blend (75/25); ▲ = PVME/PS blend (50/50); ◆ = PVME/PS blend (30/70); ▶ = PVME/PS blend (0/100)) were taken from Mokdad et al. (1996).

The stability criterion for the ternary mixture PS/PVME/CO_2 can be evaluated directly by differentiation of the Gibbs free energy (Rao and Watkins, 2000) from the following equation:

$$\begin{vmatrix} G_{xx} & G_{xy} & G_{x\tilde{v}} \\ G_{yx} & G_{yy} & G_{y\tilde{v}} \\ G_{\tilde{v}x} & G_{\tilde{v}y} & G_{\tilde{v}\tilde{v}} \end{vmatrix} > 0 \tag{70}$$

where x and y are two independent compositions variables and $\tilde{v}$ is the reduced volume of the system (Sánchez, 1982). Prediction of the stability of the pseudo binary system using the PC-SAFT EoS and SL is shown in Figure 5.7.7. Both models accurately predict a phase split upon CO_2 sorption (no change of the sign in the determinant of Eq. 70), concluding that the prediction of phase separation pressure of CO_2 made by the PC-SAFT and SL EoS is reasonably correct.

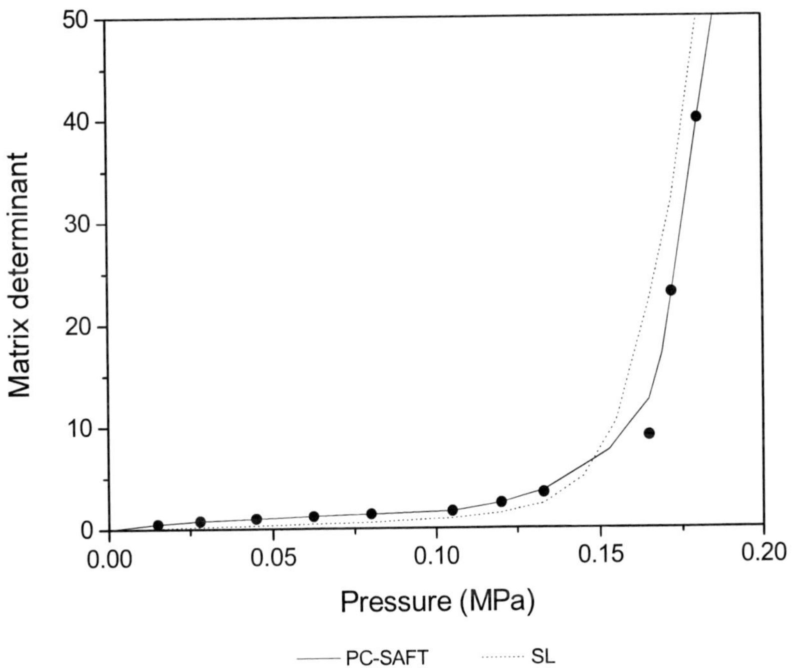

Figure 5.7.7. Stability analysis for the pseudo binary system PS/PVME blend + CO_2 using the PC-SAFT and SL EoS. Experimental data (●: 313.15 K) were taken from Rao and Watkins (2000).

Chapter 6

CONCLUSIONS

In this work, the PC-SAFT, SL and PR EoS were used to model and correlate the phase equilibria of several binary and ternary polymeric systems at different temperature and pressure conditions, by fitting just one binary interaction parameter (in many cases this parameter is dependent of temperature) in van der Waals one-type mixing rules.

Solubilities of several gases (chlorofluorocarbon, hydrochlorofluorocarbon, hydrofluorocarbon, and supercritical fluids) in PS were modeled in a range of temperatures from 293.15 to 553.15 K and pressures up to 35 MPa. Solubilities of all gases, except chlorofluorocarbon fluids, increase almost linearly with pressure; for chlorofluorocarbon gases, the solubility increases in exponential relation with the pressure. Solubilities of all gases (except N_2) in PS show a reverse behavior due to the influence of the temperature.

The solubility pressure of CO_2 in nine molten polymers [HPDE, LDPE, i-PP, p(VAc), PS, p(MMA), p(BMA), p(DMS), and PC] with different molecular weights was studied. For CO_2 + molten polymer systems, solubility of CO_2 decreases when temperature increases at constant pressure.

Cloud point isopleths were modeled in binary and ternary systems containing biodegradable polymers, copolymer and supercritical fluids. Cloud point isopleths for PLA polymer exhibited a LCST phase transition in DME, CO_2, CDFM, TFM and TFE, and a U-LCST phase transition in DFM. In these systems, the cloud point pressures increased with PLA molecular weight. Both PBS + CO_2 and PBSA + CO_2 systems have similar thermodynamic behavior; in other words, when the fluid mass percent increases, the cloud point pressure increases at a given temperature, while at a given fluid mass percent, the cloud point pressure had a little increment as the temperature increases. Cloud point isopleths also indicated a LCST phase transition for PLA in DME + CO_2 mixtures. DME is a better supercritical solvent than CO_2 for dissolving polar biodegradable polymers, although CO_2 is the favorite solvent in supercritical processes. When CO_2 was added to PLA +DME mixtures, the modeling also reported a reduction of dissolving power of the mixed solvent, probably due to the difference of polarity between DME and CO_2.

It was possible to predict PVTdata of PLAG copolymers by a group contribution method (Elvassore et al., 2002) with only the molecular structure as input. With these PVT data, EoS pure-component parameters were predicted for the three EoS. Afterwards, the cloud point isopleths were modeled by using these parameters in ten binary systems involving PLAG copolymers and SCF. Cloud point isopleths for PLAG copolymers in DME exhibited a LCST phase transition at lower pressures and a UCST phase transition at higher pressures, while in CO_2, the binary systems exhibited only a UCST phase behavior. PLAG copolymers + CDFM systems exhibited a LCST type phase behavior more clearly than the LCST type exhibited by PLAG copolymer +TFM system. DME and CDFM are the best supercritical solvents, while TFM has a less power for dissolving PLAG copolymers and CO_2 is a poor solvent for these copolymers.

Cloud pressure curves of block copolymer + supercritical CO_2 systems were simulated and correlated by PC-SAFT EoS, and a good agreement was reached when compared with the experimental data available in literature. Three block copolymer + CO_2 systems studied in this work show a LCST phase behavior; this type of phase separation is typical for polymer in supercritical fluids.

PC-SAFT and SL EoS were also used to correlate cloud point temperatures of blend miscibilities, with satisfactory predictions of the molecular weight effects on polymer blends PBD/PS, PPG/PEGE, PVME/PS, and PEO/PES and the sorption of CO_2 in PVME/PS blends. It was found that both models can satisfactorily correlate and predict the specific volumes of pure blends by introducing only one interaction parameter by adjusting the PVT experimental polymer miscibility of both the UCST and LCST. There is an influence of the molecular weight on the locations of the critical solutions temperatures. For PBD/PS and PPG/PEGE blends, when the PEGE and PS molecular weight increase, the UCST increases, while for PVME/PS and PEO/PES blends, when the PS and PEO molecular weight increase, the LCST decreases. A rigorous analysis using the PC-SAFT and SL EoS was used to know qualitatively the phase stability of the pseudo binary system PS/PVME + CO_2.

In general, the results in terms of cloud point pressure or temperature deviations indicate that the PC-SAFT model was able to correlate and predict with high accuracy the different phase behaviors provoked by the interactions of molecules of different size, such as those of polymers, copolymers, blends and supercritical fluids at high pressures and temperatures. On the other hand, since the SL EoS is based on lattice model, its results also are satisfactory for modeling the phase behavior of these systems, while the simple but efficient PR EoS obtained acceptable results.

ACKNOWLEDGMENTS

The financial support of Fundação de Amparo à Pesquisa do Estado de São Paulo, FAPESP (Brazil), through grant 05/53685-4, is gratefully acknowledged. M. Aznar is the recipient of a fellowship from Conselho Nacional de Desenvolvimento Científico e Tecnológico, CNPq (Brazil).

References

S. Alexander, Polymer Adsorption on Small Spheres. A Scaling Approach, *J. Phys. (Paris)* 38 (1977) 977.

P. Arce, Modelagem do Equilíbrio de Fases em Misturas de Dióxido de Carbono Supercrítico and Compostos presentes em Produtos Naturais. M.Sc. Thesis, (in Portuguese), State University of Campinas, Campinas, SP, Brazil, 2002.

P. Arce. Modeling and Computation of Multiphase Equilibrium of Critical Fluids and Phenomena in Polymer Solubilities in Supercritical CO_2 + Co-solvent Mixtures. D.Sc. Thesis (in Portuguese), State University of Campinas, Campinas, SP, Brazil, 2005.

P. Arce, M. Aznar, Modeling of Thermodynamic Behavior of PVT properties and Cloud Point Temperatures of Polymer Blends and Polymer Blend + Carbon Dioxide Systems using Non-Cubic Equations of State, submitted to *Polym. Eng. Sci.*

Y.C. Bae, J.J. Shim, D.S. Soane, J.M. Prausnitz, Representation of Vapor-Liquid and Liquid-Liquid Equilibria for Binary Systems containing Polymers: Applicability of an extended Flory-Huggins Equation, *J. Appl. Polym. Sci.* 47 (1993) 1193.

M. Banaszak, C.K. Chen, M. Radosz, Copolymer SAFT Equation of State. Thermodynamic Perturbation Theory extended to Heterobonded Chains, *Macromolecules* 29 (1996) 6481.

J.A. Barker, D. Henderson, Perturbation Theory and Equation-of-State for Fluids: The Square-Well Potential, *J. Chem. Phys.* 47 (1967) 2856.

E. Buhler, A.V. Dobrynin, J.M. deSimone, M. Rubinstein, Light-Scattering Study of Diblock Copolymers in Supercritical Carbon Dioxide: CO_2

Density-Induced Micellization Transition, *Macromolecules* 31 (1998) 7347.

B.H. Chang, Y.C. Bae, Molecular Thermodynamics Approach for Liquid–Liquid Equilibria of the Symmetric Polymer Blend Systems, *Chem. Eng. Sci.* 58 (2003) 2931.

W.G. Chapman, K.E. Gubbins, G. Jackson, M. Radosz, New Reference Equation of State for Associating Liquids, *Ind. Eng. Chem. Res.* 29 (1990) 1709.

S.J. Chen, I.G. Economou, M. Radosz, Density-Tuned Polyolefin Phase-Equilibria 2. Multicomponent Solutions of Alternating Poly(Ethylene Propylene) in Subcritical and Supercritical Olefins - Experiment and SAFT Model, *Macromolecules* 25 (1992) 4987.

C.M. Colina, C.K. Hall, K.E. Gubbins. Phase Behavior of PVAC-PTAN Block Copolymer in Supercritical Carbon Dioxide using SAFT, *Fluid Phase Equilib.* 194-197 (2002) 553.

S.E. Conway, H.-S. Byun, M.A. McHugh, J.D. Wang, F.S. Mandel, Poly(lactide-co-glycolide) Solution Behavior in Supercritical CO_2, CHF_3, and $CHClF_2$, *J. Appl. Polym. Sci.* 80 (2001) 1155.

C. Dariva, J.V. Oliveira, F.W. Tavares, J.C. Pinto, Phase Equilibria of Polypropylene samples with Hydrocarbon Solvents at High Pressures, *J. Appl. Polym. Sci.* 81 (2001) 3044.

P.K. Davis, G.D. Lundy, J.E. Palamara, J.L. Duda, R.P. Danner, New Pressure-Decay Techniques to Study Gas Sorption and Diffusion in Polymers at Elevated Pressures, *Ind. Eng. Chem. Res.* 43 (2004) 1537.

P.G. de Gennes, Conformations of Polymers Attached to an Interface, *Macromolecules* 13 (1980) 1069.

DIPPR Information and Data Evaluation Manager. Version 1.2.0, 2000.

H.J. Domb, J. Kost, D.M. Wiseman, Handbook of Biodegradable Polymers, Drug Targeting and Delivery Series, vol. 7, Harwood Academic Publishers, Amsterdam, 1997.

P.L. Durrill, R.G. Griskey, Diffusion and Solution of Gases in Thermally Softened or Molten Polymers. Part 1. Development of Technique and Determination of Data, *AIChE J.* 12 (1966) 1147.

N. Elvassore, A. Bertucco, M. Fermeglia, Phase-Equilibria Calculation by Group-Contribution Perturbed-Hard-Sphere-Chain Equation of State, *AIChE J.* 48 (2002) 359.

R.R. Edwards, Y. Tao, S. Xu, P.S. Well, K.S. Yun, J.F. Parcher, Chromatographic Investigation of CO_2 - polymer Interactions at Near-critical Conditions, *J. Phys. Chem. B.* 102 (1998) 1287.

O.A. Evers, J.H.M.M. Scheutjens, G.J. Fleer, Statistical Thermodynamics of Block Copolymer Adsorption. 1. Formulation of the Model and Results for the Adsorbed Layer Structure, *Macromolecules* 23 (1990) 5221.

M.J. Fasolka, A.M. Mayes, Block Copolymer Thin Films: Physics and Applications, *Annu. Rev. Mater. Res.* 31 (2001) 323.

G.J. Fleer, M.A. Cohen Stuart, J.M. Scheutjens, T. Cosgrove, B. Vincent, Polymers at Interfaces, Chapman and Hall, London, 1993.

B. Folie, M. Radosz, Phase Equilibria on High-pressure Polyethylene Technology, *Ind. Eng. Chem. Res.* 34 (1995) 1501.

A. Garg, E. Gulari, C.W. Manke, Thermodynamics of Polymer Melts Swollen with Supercritical Gases, *Macromolecules* 27 (1994) 5643.

R.A. Gorski, R.B. Ramsey, K.T. Dishart, Physical properties of blowing agent polymer systems. I. Solubility of fluorocarbon blowing agents in thermoplastic resins, *J. Cell. Plast.* 22 (1986) 21.

J. Gross, G. Sadowski, Application of Perturbation Theory to a Hard-chain Reference Fluid: An Equation of State for Square-well Chains, *Fluid Phase Equilib.* 168 (2000) 183.

J. Gross, G. Sadowski, Perturbed-Chain SAFT: An Equation of State based on a Perturbation Theory for Chain Molecules, *Ind. Eng. Chem. Res.* 40 (2001) 1244.

J. Gross, G. Sadowski, Application of the Perturbed-Chain SAFT Equation of State to Associating Systems, *Ind. Eng. Chem. Res.* 41 (2001a) 5510.

J. Gross, G. Sadowski, Modeling Polymer Systems using the Perturbed-Chain Statistical Associating Fluid Theory Equation of State, *Ind. Eng. Chem. Res.* 41 (2002) 1084.

J. Gross, O. Spuhl, F. Tumakaka, G. Sadowski, Modeling Copolymer Systems using the Perturbed-Chain SAFT, *Ind. Eng. Chem. Res.* 42 (2003) 1266.

G. Hadziioannou, S. Patel, S. Granick, J. Tirrell, Forces between Surfaces of Block Copolymers Adsorbed on Mica, *J. Am. Chem. Soc.* 108 (1986) 2869.

E. Helfand, Z.R. Wasserman, Block Copolymer Theory. 4. Narrow Interphase Approximation, *Macromolecules* 9 (6) (1976) 879.

A. Hoshino, Y. Isono, Degradation of Aliphatic Polyester Films by commercial avaliable Lipases with special reference to rapid and complete Degradation of Poly(L-lactide) Film by Lipase PL derived from Alcaligenes sp, *Biodegration* 13 (2002) 141.

Y-L. Hsiao, E.E. Mauri, J.M. deSimone, S. Mawson, K.P. Johnston, Dispersion Polymerization of Methyl Methacrylate Stabilized with

Poly(1,1-dihydroperfluorooctyl acrylate) in Supercritical Carbon Dioxide, *Macromolecules* 28 (1995) 8159.

S.H. Huang, M. Radosz, Equation of State for Small, Large, Polydisperse and Association Molecules, *Ind. Chem. Eng. Res.* 29 (1990) 2284.

S.H. Huang, M. Radosz, Equation of State for Small Large, Polydisperse and Associating Molecules: Extention to Fluid Mixtures, *Ind. Chem. Eng. Res.* 30 (1991) 1994.

M. Imaizumi, R. Fujihira, J. Suzuki, K. Yoshikawa, R. Ishioka, M. Takahashi, A Study of Biodegradable Foam of Poly(butylenes succinate) (PBS). The Relationship among Branching Factor, Rheological Properties and Processability of Direct Extrusion Gas Foaming for PBS, *J. Jpn. Soc. Polym. Proces.* 11 (1999) 432.

S. Jacobsen, H.G. Fritz, R. Jerome, Polylactide, *Polym. Eng. Sci.* 39 (1999) 1311.

R.A. Jain, The Manufacturing Techniques of various Drug Loaded Biodegradable Poly(lactide-co-glycolide) (PLGA) Devices, *Biomaterials* 21 (2000) 2475.

A. Jónasson, O. Persson, Aa. Fredenslund, High Pressure Solubility of Carbon Dioxide and Carbon Monoxide in Dimethyl Ether, *J. Chem. Eng. Data* 40 (1995) 296.

S. Karlsson, A.C. Albertson, Biodegradable Polymers and Environmental Interaction, *Polym. Eng. Sci.* 38 (1998) 1251.

J.U. Keller, H. Rave, R. Staudt, Measurement of Gas Absorption in a Swelling Polymeric Material by a combined Gravimetric-dynamic Method, *Macromol. Chem. Phys.* 200 (1999) 2269.

M. Kinzl, G. Luft, H. Adidharma, M. Radosz, SAFT Modeling of Inert-gas Effects on the Cloud-point Pressures in Ethylene Copolymerization Systems: Poly(ethylene - co - vinyl acetate) + Vinyl acetate + Ethylene and Poly(ethylene - co - hexene-1) + Hexene-1 + Ethylene with Carbon Dioxide, Nitrogen, *Ind. Eng. Chem. Res.* 39 (2000) 541.

R. Koningsveld, W.H. Stockmayer, E. Nies, Polymer Phase Diagrams. A Textbook, Oxford University Press, New York, 2001.

Y.M. Kuk, B.C. Lee, Y.W. Lee, J.S. Lim, Phase Behavior of Biodegradable Polymers in Dimethyl Ether and Dimethyl Ether + Carbon Dioxide, *J. Chem. Eng. Data* 46 (2001) 1344.

Y.-M. Kuk, B.-C. Lee, Y.-W. Lee, J.S. Lim, High-pressure Phase Behavior of Poly(D,L-lactide) in Chlorodifluoromethane, Difluoromethane, Trifluoromethane, and 1,1,1,2-Tetrafluoroethane, *J. Chem. Eng. Data* 47 (2002) 575.

R.K. Kulkarni, S.G. Moore, A.F. Higyeli, F. Leonard, Biodegradable Poly(Lactic Acid) Polymers, *J. Biomed. Mater. Res.* 5 (1971) 169.

T. Laursen, P. Rasmussen, S.I. Andersen, VLE and VLLE Measurements of Dimethyl Ether containing Systems, *J. Chem. Eng. Data* 47 (2002) 198.

J.M. Lee, B.-C. Lee, S.-H. Lee, Cloud Points of Biodegradable Polymers in Compressed Liquid and Supercritical Chlorodifluoromethane, *J. Chem. Eng. Data* 45 (2000) 851.

H. Liu, Y. Hu, Molecular Thermodynamic Theory for Polymer Systems. II. Equation of State for Chain Fluids, *Fluid Phase Equilib.* 122 (1996) 75.

T.P. Lodge, Block Copolymers: Past Successes and Future Challenges, *Macromol. Chem. Phys.* 204 (2) (2003) 265.

V. Louli, D. Tassios, Vapor-liquid Equilibrium in Polymer-solvent Systems with a Cubic Equations of State, *Fluid Phase Equilib.* 168 (2000) 165.

J.L. Lundberg, M.B. Wilk, M.J. Huyett, Sorption Studies using Automation and Computation, *Ind. Eng. Chem. Fundam.* 2 (1963) 37.

S. Mawson, K.P. Johnston, J.R. Combes, J.R. deSimone, Formation of Poly(1,1,2,2-tetrahydroperfluorodecyl acrylate) Submicron Fibers and Particles from Supercritical Carbon Dioxide Solutions, *Macromolecules* 28 (1995) 3182.

J.M. Mayer, D.L. Kaplan, Biodegradable Materials: balancing Degradability and Performance, *Trends Polym. Sci.* 2 (1994) 227.

M.A. McHugh, V.J. Krukonis, Supercritical Fluid Extraction. Principles and Practice, Butterworths, Stoneham, MA, 1994.

J.C. Middleton, A.J. Tipton, Synthetic Biodegradable Polymer as Orthopedic Devices, *Biomaterials* 21 (2000) 2335.

A. Mokdad, A. Dubault, L. Monnerie, Sorption and Diffusion of Carbon Dioxide in single-phase Polystyrene/Poly(vinylmethylether) Blends, *J. Polym. Sci.: Part B: Polym. Phys.* 34 (1996) 2723.

P.M. Ndiaye, C. Dariva, J.V. Oliveira, F.W. Tavares, Phase Behavior of Isotactic Polypropylene/c4-Solvents at High Pressure. Experimental Data and SAFT Modeling, *J. Supercritical Fluids*, 21 (2001) 93.

D.M. Newitt, K.E. Weale, Solutions and Diffusion of Gases in Polystyrene at High Pressures, *J. Chem. Soc. London* Oct. (1948) 1541.

V.G. Niesen, V.F. Yesavage, Application of a Maximum Likelihood Method using Implicit Constraints to determine Equation of State Parameters from Binary Phase Behavior Data, *Fluid Phase Equilib.* 50 (1989) 249.

R. O'Lenick, X.J. Li, Y.C. Chiew, *Correlation Functions of Hard-sphere Chain Mixtures: Integral Equation Theory and Simulation Results, Mol. Phys.* 86 (1995) 1123.

M.L. O'Neill, Q. Cao., C.M. Fang, K.P. Johnston, S.P. Wilkinson, C.D. Smith, J.L. Kerschner, S.H. Jureller, Solubility of Homopolymers and Copolymers in Carbon Dioxide, *Ind. Eng. Chem. Res.* 37 (1998) 3067.

C. Pan, M. Radosz, Copolymer SAFT Modeling of Phase Behavior in Hydrocarbon-chain Solutions: Alkane Oligomers, Polyethylene, Poly(ethylene-co-olefin-1), Polystyrene, and Poly(ethylene-co-styrene), *Ind. Eng. Chem. Res.* 37 (1998) 3169.

D.R. Paul, S. Newman, S., Eds.; Polymer Blends; Academic Press: New York, 1978.

D.Y. Peng, D.B. Robinson, A New Two Constant Equation of State, *Ind. Eng. Chem. Fundam.* 15 (1976) 59.

C.J. Peng, H.L. Liu, Y. Hu, Gas Solubilities in Molten Polymers based on an Equation of State, *Chem. Eng. Sci.* 56 (2000) 6967.

D.S. Pope, I.C. Sánchez, W.J. Koros, G.K. Fleming, Statistical Thermodynamic Interpretation of Sorption/dilation Behavior of Gases in Silicone Rubber, *Macromolecules* 24 (1991) 1779.

V.S. Ramachandra Rao, J.J. Watkins, Phase Separation in Polystyrene - Poly(vinyl methyl ether) Blends dilated with Compressed Carbon Dioxide, *Macromolecules* 33 (2000) 5143.

P.A. Rodgers, Pressure-Volume-Temperature Relationships for Polymer Liquids: A Review of Equations of State and their Characteristic Parameters for 56 Polymers, *J. Appl. Polym. Sci.* 48 (1993) 1061.

S. Rostami, D.J. Walsh, Simulation of Upper and Lower Critical Phase Diagrams for Polymer Mixtures at various Pressures, *Macromolecules* 18 (1985) 1228.

E. Sada, H. Kumazawa, H. Yakushiji, Y. Bamba, K. Sakata, S.T. Wang, Sorption and Diffusion of Gases in Glass Polymer, *Ind. Eng. Chem. Res.* 26 (1987) 433.

I.C. Sánchez, R.H. Lacombe, An Elementary Molecular Theory of Classical Fluids. Pure Fluids, *J. Phys. Chem.* 80 (1976) 2352.

I.C. Sánchez, R.H. Lacombe, Statistical Thermodynamics of Polymer Solutions, *Macromolecules* 11 (1978) 1145.

I.C. Sánchez, In: Polymer Compatibility and Incompatibility Principles and Practices, K. Sole (ed.), MIMIM Press Symposium Series Vol. 2, Hardwood Academic Publishers, Chur, Switzerland, 1982.

Y. Sato, M. Yurugi, K. Fujiwara, S. Takishima, H. Masuoka, Solubilities of Carbon Dioxide and Nitrogen in Polystyrene under High Temperature and Pressure, *Fluid Phase Equilib.* 125 (1996) 129.

Y. Sato, K. Fujiwara, T. Takikawa, S. Takishima, H. Masuoka, Solubilities and Diffusion Coefficients of Carbon Dioxide and Nitrogen in Polypropylene, High-density Polyethylene, and Polystyrene under High Pressures and Temperatures, *Fluid Phase Equilib.* 162 (1999) 261.

Y. Sato, T. Iketani, S. Takishima, H. Masuoka, Solubility of Hydrofluorocarbon (HFC-134a, HFC-152a) and Hydrochlorofluorocarbon (HCFC-142b) Blowing Agents in Polystyrene, *Polym. Eng. Sci.* 40 (2000) 1369.

Y. Sato, T. Takikawa, A. Sorakubo, S. Takishima, H. Masuoka, M. Imaizumi, Solubility and Diffusion Coefficients of Carbon Dioxide in Biodegradable Polymers, Ind. *Eng. Chem. Res.* 39 (2000a) 4813.

Y. Sato, T. Takikawa, S. Takishima, H. Masuoka. Solubilities and Diffusion Coefficients of Carbon Dioxide in Poly(vinyl acetate) and Polystyrene, *J. Supercritical Fluids* 19 (2001) 187.

Y. Sato, M. Wang, S. Takishima, H. Masuoka, T. Watanabe, Y. Fukasawa, Solubility of Butane and Isobutane in Molten Polypropylene and Polystyrene, *Polym. Eng. Sci.* 44 (2004) 2083.

K.P. Shukla, W.G. Chapman, SAFT Equation of State for Fluid Mixtures of Hard Chain Copolymers, *Mol. Phys.* 91 (1997) 1075.

Y. Song, S.M. Lambert, J.M. Prausnitz, A Perturbed Hard-Sphere-Chain Equation of State for Normal Fluids and Polymers, *Ind. Eng. Chem. Res.* 33 (1994) 1047.

Y. Song, T. Hino, S.M. Lambert, J.M. Prausnitz, Liquid-liquid Equilibria for Polymer Solutions and Blends, including Copolymers, *Fluid Phase Equilib.* 117 (1996) 69.

T. Spyriouni, I.G. Economou, Evaluation of SAFT and PC-SAFT Models for the Description of Homo- and Co- Polymer Solution Phase Equilibria, *Polymer* 46 (2005) 10772.

R. Stockfleth, R. Dohrn, An Algorithm for calculating Critical Points in Multicomponent Mixtures which can easily be implemented in existing Programs to calculate Phase Equilibria, *Fluid Phase Equilib.* 145 (1998) 43.

L. Stragevitch, S.G. d'Avila, Application of a Generalized Maximum Likelihood Method in the reduction of Multicomponent Liquid–liquid Equilibrium Data, *Braz. J. Chem. Eng.* 14 (1997) 41.

K. Takahashi, T. Kyu, Q. Trancong, O. Yano, T. Soen, Phase-separation in Mixtures of Poly(ethylene glycol monomethylether) and Poly(propylene glycol) Oligomers, *J. Polym. Sci. B. Polym. Phys.* 29 (1991) 1419.

S. Takishima, K. Nakamura, M. Sasaki, H. Masuoka, Dilation and Solubility in Carbon Dioxide + Poly(vinyl acetate) System at High Pressure, *Sekiyu Gakkaishi*. 33 (1990) 332.

M. Tang, T.B. Du, Y.P. Chen, Sorption and Diffusion of Supercritical Carbon Dioxide in Polycarbonate, *J. Supercritical Fluids* 28 (2004) 207.

T. Teeraphatpornchai, T. Nakajima-Kambe, Y. Shigeno-Akutsu, M. Nakayama, N. Nomura, T. Nakahara, H. Uchiyama, Isolation and Characterization of a Bacterium that degrades various Polyester-based Biodegradable Plastics, *Biotechnol. Lett.* 25 (2003) 23.

J.W. Tom, P.G. Debenedetti, Formation of Bioerodible Polymeric Microspheres and Microparticles by rapid expansion of Supercritical Solutions, *Biotech. Prog.* 7 (1991) 403.

F. Tumakaka, J. Gross, G. Sadowski, Modeling of Polymer Phase Equilibria using the Perturbed-Chain SAFT, *Fluid Phase Equilib.* 194–197 (2002) 541.

D. Urban, K. Takamura. Polymer Dispersions and Their Industrial Applications. Wiley-VCH Verlag GmbH, Weinheim, 2002.

L.A. Utracki, L. A., Polymer Alloys and Blends, Hanser Publishers, Munich, 1989.

L.A. Utracki, Polymer Blends Handbook, Kluwer Academic Publishers, vol. I., The Netherlands, 2002.

N. von Solms, J.K. Nielsen, O. Hassager, A. Rubin, A.Y. Dandekar, S.I. Andersen, E.H.J. Stenby, Direct Measurement of Gas Solubilities in Polymers with a High-pressure Microbalance, *J. Appl. Polym. Sci.* 91 (2004) 1476.

E.P. Voutsas, G.P. Pappa, C.J. Boukouvalas, K. Magoulas, D.P. Tassios, Miscibility in Binary Polymer Blends: Correlation and Prediction, *Ind. Eng. Chem. Res.* 43 (2004) 1312.

D.J. Walsh, G.T. Dee, J.L. Halary, J.M. Ubiche, M. Millequant, J. Lesec, L. Monnerie, The Application of Equation-of-state Theories to narrow Molecular-weight Distribution Mixtures of Polystyrene and Poly(vinyl methyl ether), *Macromolecules* 22 (1989) 3395.

D.J. Walsh, The Role of Specific Interactions in Polymer Miscibility. In Integration of Fundamental Polymer Science and Technology, Kleintjens L.A., Lemstra P.S., Eds.: Elsevier. New York, 1986.

N.H. Wang, S. Takishima, H. Masuoka, Solubility Measurements of Gas in Polymer by the Piezoelectric-quartz Sorption Method at High-pressures and its Correlation. Carbon Dioxide + Poly(vinyl acetate) and Carbon dioxide + Poly(butyl methacrylate) Systems, *Kagaku Kogaku Ronbunshu* 16 (1990) 931.

N.H. Wang, S. Takishima, H. Masuoka, Measurement and Correlation of Solubility of High Pressure Gas in a Polymer by Piezoelectric Quartz Sorption - CO_2 + PVAc and CO_2 + PBME Systems, *Int. Chem. Eng. (Japan)* 34 (1994) 255.

S. Wang, C. Peng, K. Li, J. Wang, J. Shi, H.J. Liu, Measurement of Solubilities of CO_2 in Polymers by Quartz Crystal Microbalance, *J. Chem. Ind. Eng. (China)* 54 (2003) 141.

Y. Wang. Molecular Modeling applied to CO_2-soluble Molecules and Confined Fluids. Ph.D. Thesis, University of Pittsburgh, Pittsburgh, PA, USA, 2006.

C. Witt, K. Mäder, T. Kissel, The Degradation, Swelling and Erosion Properties of Biodegradable Implants prepared by Extrusion or Compression Moulding of Poly(lactide-co-glycolide) and ABA Triblock Copolymers, *Biomaterials* 21 (2000) 931.

INDEX

A

accuracy, 32, 47, 49, 52, 96, 101, 109
acetate, 42, 86, 116, 119, 120
acid, 5
acidic, 84, 85
acrylate, 86, 116, 117
acrylonitrile, 7
adipate, 5, 53
adsorption, 7
agent, 4, 40, 115
alkanes, 11, 17
alternative, 4, 5, 14, 87
amorphous, 7
application, 4
arithmetic, 12, 19
asymmetry, 7
atoms, 4, 84

B

biocompatibility, 5
biodegradable, 4, 5, 7, 21, 53, 56, 76, 107
biodegradable materials, 5
biomass, 5
biomedical applications, 5
bisphenol, 42
blends, 1, 3, 6, 7, 23, 95, 96, 97, 98, 99, 100, 101, 102, 103, 104, 108, 109
blocks, 6
blowing agent, 4, 40, 115
BMA, 42, 43, 44, 45, 50, 51, 107
bonding, 9, 14, 85, 87
butadiene, 7, 95
butane, 27
butyl methacrylate, 42, 121

C

candidates, 6
carbon, 3, 4, 42, 113, 114, 116, 117, 118, 119, 120
carbon dioxide, 4, 42
chemical structures, 92
chloride, 4
chlorofluorocarbons, 4, 27
clusters, 9
coatings, 4, 6
commodity, 5
components, 1, 3, 13, 22, 26
composition, 3, 14, 21, 75, 87
compounds, vi, 1, 3, 4, 5, 26
compressibility, 16

concentration, 59, 63, 66, 67, 70, 72, 74, 75, 89, 92, 94, 97, 102
copolymer, 1, 3, 5, 6, 7, 9, 12, 13, 14, 16, 20, 21, 26, 27, 53, 56, 58, 76, 77, 78, 79, 80, 81, 82, 83, 84, 86, 87, 88, 89, 90, 91, 92, 94, 107, 108, 109
correlation, 40, 42, 65, 98, 99, 100, 101
cosmetics, 1
cost-effective, 6
covalent, 9
covalent bond, 9
critical points, 7, 73
critical temperature, 3, 32, 58
crystals, 6

D

data set, 42
degradation, 5
delivery, 5, 6
density, 1, 3, 12, 16, 17, 27, 42, 82, 88, 93, 119
deviation, 25, 35, 42, 44, 49, 52, 58, 73, 77, 80
devolatilization, 4
differentiation, 22, 23, 104
dilation, 118
dipole, 58, 60, 85, 102
dipole moment, 58, 60, 85
dispersion, 9, 11, 13, 15, 16
distribution, 10, 11, 12, 14, 15, 16
distribution function, 10, 11, 12, 15, 16
drug delivery, 5, 6
drug delivery systems, 5

E

energy, 10, 13, 14, 16, 20, 24, 32, 58, 82, 104
entropy, 88
environment, 4
enzymes, 4
equilibrium, vi, 9, 21, 23, 42, 58, 77, 87, 88, 90, 94, 102
ester, 85
estimating, 26, 86
ether, 116, 117
ethylene, 27, 86, 95, 116, 118, 119
ethylene glycol, 95, 119
ethylene oxide, 86, 95
experimental condition, 92
extrusion, 4, 5

F

fermentation, 5
financial support, 111
fixation, 5
fluid, 1, 11, 12, 16, 21, 28, 32, 33, 35, 46, 58, 59, 60, 68, 70, 71, 72, 73, 76, 77, 81, 84, 86, 87, 88, 93, 95, 96, 107
fluorine, 4
fluorine atoms, 4
foams, 4
fracture, 5
free energy, 24, 104
free volume, 7, 88

G

gas, 4, 5, 17, 21, 27, 28, 32, 40, 42, 46, 107, 116
gas phase, 21
Gibbs free energy, 24, 104
glycol, 95, 119
groups, 26, 85

H

HDPE, 42, 44, 46
helium, 32
high pressure, 3, 42, 76, 77, 109
high temperature, 1, 3
homopolymers, 6, 13

HSC, 86
hydro, 4
hydrocarbons, 4
hydrogen, 9, 27, 84, 85
hydrogen atoms, 84
hydrogen bonds, 9

I

in vitro, 5
in vivo, 5
industrial, 1, 3, 5, 6, 7
industrial application, 1, 3
industrial production, 7
industrial sectors, 1
inert, 4
interaction, 1, 3, 6, 7, 9, 12, 18, 31, 33, 35, 40, 42, 44, 45, 52, 57, 58, 73, 74, 77, 80, 82, 85, 88, 90, 96, 98, 100, 102, 107, 108, 109
interface, 93
intermolecular, 10
interval, 58, 76
isotherms, 35, 46, 47, 49, 50, 60, 63, 64, 66, 67, 68, 69, 70, 72, 92, 93, 102, 103, 104

L

lactic acid, 5
lattice, 17, 109
Life Cycle Assessment, iii
likelihood, 25, 31, 58, 77
linear, 32, 35, 37, 61, 70, 71, 102
liquid phase, 21, 23
L-lactide, 5, 53, 77, 78, 81, 82, 83, 84, 85, 115, 116
location, 5, 7
low molecular weight, 1, 3, 7, 25
low temperatures, 67
low-density, 42
lubricants, 6

M

macromolecules, 9
Maximum Likelihood, 117, 119
mechanical properties, 5
membranes, 1
methane, 27
methyl group, 89, 90
methyl methacrylate, 42
microbial, 5
mixing, 12, 16, 18, 19, 22, 23, 73, 107
MMA, 42, 43, 44, 45, 49, 50, 107
models, 1, 3, 4, 9, 14, 26, 27, 32, 42, 44, 49, 58, 61, 62, 63, 65, 67, 68, 71, 73, 74, 77, 84, 85, 87, 88, 92, 96, 97, 98, 100, 104, 108, 109
molar ratio, 78, 81, 82, 83, 84, 85
molar volume, 18, 22, 25, 27
mole, 12, 21, 22, 23, 24, 73, 77, 81, 84, 87, 89, 92, 94
molecular structure, 27, 58, 77, 108
molecular weight, vi, 1, 3, 4, 7, 13, 25, 26, 27, 42, 62, 65, 66, 68, 75, 76, 78, 90, 96, 97, 98, 100, 101, 107, 108
molecules, 1, 3, 7, 9, 12, 14, 58, 77, 109
monomer, 4, 5, 13, 14, 52, 78, 90
multiplication, 22

N

natural, 88
nitrogen, 4, 27

O

observations, 46
orientation, 99
oscillation, 6
oxide, 86, 95
oxygen, 32
ozone, 4

P

paints, 1, 6
parameter, 12, 13, 16, 18, 20, 32, 47, 52, 57, 62, 88, 107, 108
peptide, 5
perturbation, 11, 12, 16
perturbation theory, 12
phase boundaries, 3, 5, 7
phase diagram, 7, 8, 93, 98
phase transitions, 7
photonic, 6
photonic crystals, 6
physical interaction, 1, 3
physical properties, 6, 43
plastics, 4, 5
PLGA, 116
polarity, 75, 85, 102, 108
poly(glycolide), 5
poly(methyl methacrylate), 42
polyester, 5
polyethylene, 42
polymer blends, 6, 7, 23, 95, 97, 108
polymer chains, 6, 99
polymer properties, 6
polymer solubility, 7
polymer solutions, 1, 3
polymer systems, 43, 45, 46, 52, 107, 115
polymerization, 4
polypropylene, 42
polystyrene, 7, 27, 42, 95
polyvinyl chloride, 4
poor, 108
power, 11, 16, 17, 108
prediction, 73, 74, 104
production, 4, 5, 7
property, vi, 22
propylene, 86, 119
proteins, 5
pseudo, 42, 102, 104, 105, 108
PVAc, 121

Q

quadrupole, 82, 85
quartz, 120

R

radial distribution, 10, 11, 15, 16
random, 13, 14
range, 1, 5, 26, 27, 32, 42, 58, 76, 77, 95, 97, 107
raw material, 5
regular, 13
relationship, 6, 21, 23, 37, 60
research and development, ix, 1
resins, 115
rubber, 7, 118

S

saturation, 58, 60, 62, 63, 65, 67, 68, 75, 82, 85
separation, 6, 88, 92, 105, 108, 119
series, 11, 12, 16, 17
sign, 104
siloxane, 7, 42
sites, 17, 19
solubility, 4, 7, 21, 27, 32, 35, 37, 40, 42, 44, 46, 47, 49, 71, 88, 89, 90, 102, 107
solvent, ix, 1, 3, 6, 7, 8, 9, 25, 31, 52, 53, 55, 57, 58, 74, 75, 76, 77, 82, 84, 85, 88, 108, 113, 117
solvent molecules, 1, 3, 7
sorption, 102, 103, 104, 108
sorption isotherms, 102, 103, 104
spheres, 9, 10
stability, 6, 104, 108
storage, 6
styrene, 7, 118
styrene-butadiene rubber (SBR), 7
substances, 1, 3

substitution, 89, 90
suffering, 59
supercritical, vi, ix, 1, 3, 21, 27, 28, 31, 42, 53, 55, 57, 62, 76, 77, 86, 87, 88, 90, 91, 92, 93, 94, 107, 108, 109
supercritical fluids, vi, ix, 1, 3, 27, 28, 42, 53, 88, 92, 107, 108, 109
surgical, 5
suspensions, 6

T

technology, 1
temperature dependence, 7
tension, 89
TFE, 53, 54, 55, 57, 68, 69, 70, 107
thermodynamic, 1, 3, 4, 6, 12, 14, 21, 23, 25, 27, 32, 42, 51, 52, 53, 58, 63, 68, 73, 74, 77, 85, 95, 96, 98, 101, 107
thermodynamic properties, 6
thermoplastic, 115
toxic, 4
toxicological, 5
transition, 7, 58, 60, 90, 92, 107, 108

V

values, 7, 23, 26, 50, 51, 73, 77
vapor, 25, 26, 27, 73, 77
variables, 15, 35, 77, 92, 104
vector, 23
viscoelastic properties, 5
viscosity, 78
visualization, 85

W

workers, 26